eYouth

Balancing Between Opportunities and Risks

P.I.E. Peter Lang

Bruxelles · Bern · Berlin · Frankfurt am Main · New York · Oxford · Wien

Michel Walrave, Wannes Heirman, Sara Mels,
Christiane Timmerman & Heidi Vandebosch (eds.)

eYouth

Balancing Between Opportunities and Risks

Cover picture: from iStockphoto.com © by A-Digit.

Éditions scientifiques internationales
Brussels, 2012
1 avenue Maurice, B-1050 Brussels, Belgium
www.peterlang.com; info@peterlang.com

Printed in Germany

ISBN 978-90-5201-784-6
D/2012/5678/11

Library of Congress Cataloging-in-Publication Data
E-youth : balancing between opportunities and risks / Michel Walrave ... [et a.] (eds.). p. cm. Papers presented at a conference organised by UCSIA and MIOS University of Antwerp. ISBN 978-90-5201-784-6 (alk. paper)
1. Internet and teenagers--Congresses. 2. Internet and children--Congresses. 3. Internet--Safety measures--Congresses. 4. Technology and youth--Congresses. 5. Technology and children--Congresses. 6. Cyberbullying--Congresses. I. Walrave, Michel. HQ799.2.I5E15 2011
004.67'80835--dc23 2011045111

CIP also available from the British Library, GB

"Die Deutsche Nationalbibliothek" lists this publication in the "Deutsche Nationalbibliografie"; detailed bibliographic data is available on the Internet at <http://dnb.d-nb.de>.

Table of Contents

Preface

Most children and adolescents of the present generation have grown up in households where information and communication technologies are readily available. The omnipresence of new media within modern households has certainly given teenagers plenty of advantages and opportunities with respect to education, entertainment and social contact. Despite the obvious enrichment of children's and adolescents' lives through ICT, it is not hard to think of potential pitfalls in the use of these technologies, such as aggressivity in online contacts and confrontation with age-restricted and possibly harmful content.

This duality, the balance between online opportunities and risks, was the central theme of a two day international multidisciplinary conference organised by UCSIA and MIOS University of Antwerp. The conference, which was entitled "eYouth: Balancing between opportunities and risks", attracted a broad multidisciplinary audience. Eighty participants were selected to present their papers at the conference following a double-blind peer review process. Around two hundred participants from dozens of countries attended the conference.

With this publication, the conference organisers wish to share important contributions of the eYouth conference with both the academic and the professional world. In this way, they want to contribute to the ongoing debate regarding both the opportunities and risks faced by children and teenagers on the Internet.

This book is divided into four parts, preceded by a general introduction in which we wish to inform the reader of the latest research in the field of the use of new technologies and the Internet by children and teenagers. The introduction focuses on the current academic debate dealing with the (im)balance between online content- and contact-related opportunities and risks. First, a typology of several concrete risks and opportunities is sketched. This is followed by a review of research results and discussions of some topical issues.

Before elaborating the specific benefits and threats related to the use of ICT, this book focuses on the fact that the term "eYouth" is not to be considered as a homogeneous group, one generation of digital natives who share the same access to technology and have the same usage patterns. In fact, youngsters differ in the way in which they integrate ICT in their daily lives. That is why the first part of the book is devoted to the issue of online *access*. If young people do not have access or decide not to use ICT, they are not in the position to benefit from the opportunities provided by the new media. This first part analyses ICT ownership, differentiated usage and varying competence among teenagers.

Part two and part three of the book are devoted to two very important reasons why so many children and teenagers use ICT. Firstly, there is a huge amount of interesting content available for teenagers online and, what is more, due to certain features of computer-mediated technologies they are able to become content producers themselves. Secondly, existing surveys have clearly pointed out that the communication and contact possibilities of ICT are the main reason why many teenagers use these media so intensely. Therefore, both opportunities and risks with respect to online content and contact are investigated.

The last part of the book is devoted to the question of empowerment and protection. Although the benefits of ICT are numerous, there are certain risks and perils that under-aged people may encounter online. That is why the role of parents in supporting youngsters, in order to maximise benefits and minimise risks, is investigated. Moreover, the policy initiatives taken in several parts of the world to protect minors online, are also analysed.

Below we discuss the underlying structure of each part of the book in more detail.

Online access. Chapter one demonstrates the large variety of new media technologies that are appropriated by Flemish teenagers. The devices, their properties and the contexts in which they are used play an important role in how these media are received and consumed. The second chapter addresses the question, how digital inequalities persist among young Belgian people, notwithstanding the fact that only a few of them are completely disconnected from the Internet at this date in time.

Online content. Chapter three investigates the social meaning of young people's creative use of digital technologies and the Internet in particular. The authors explore how online creativity and sociability are intertwined and which motivations underlie teenagers' decisions to become creative online producers. Chapter four sheds new light on the political participation of young adults by using unique behavioural data that show the age distribution of online political participation. The study relies on two cases: posting messages on online political forums and problem reporting by email to local authorities. On the other hand, online content also includes news sources that young people can consult. How does online news compete with information presented in traditional mass media? This question is addressed in chapter five. The author discusses the extent to which children are interested in the news and the role that the Internet with its many online news resources plays in this regard. The results of this study indicate that news from traditional media is still more popular than online sources with young people and that the advantages of online news – such as flexible strings of news, continuous updates and a nearly limitless supply of information – are insufficiently known and used by youngsters. The final chapter of this second part deals with parental mediation in digital gaming, taken as a mainstream activity of children and adolescents living in contemporary society.

Online contact. Chapters seven and eight both tackle an emerging ICT-related social problem among youngsters, often referred to as cyberbullying. Chapter seven outlines the findings of a German study on broadening the understanding of the behaviour of cyberbullying by analysing the influence of the perceived feelings (one's own and those of others) towards cyberbullying incidents. Chapter eight takes a different angle and examines the students' perspective on which actions should be taken to prevent cyberbullying in schools. Besides these risks, opportunities for contact are also dealt with in this section of the volume. Specifically, the possibilities to create an online profile and to get in touch with other youngsters online offer an important opportunity. Chapter nine reports the findings of a Spanish study dealing with the question of how Spanish teenagers use a social network site, *Fotolog*, for socialising love questions. In order to assess whether self-presentation on this social networking site is informed by a conventional or a transformative model, 400 entries on *Fotolog* were analysed.

Empowerment and protection. The study in chapter ten used data from the Flash Eurobarometer 248 to construct a typology of parents, differentiating five types of parents based on the strategies they use for mediating their children's Internet use. By developing a classification of countries, a cross-national comparison is made about which types of parental mediation predominate in each country. Chapter eleven presents an analysis of the applicability of the safeguards contained in the relevant articles of the European Convention on Human Rights and of the requirements that need to be met to ensure an optimal protection for minors from harmful digital content. Chapter twelve reviews the efforts that have been made by China, Japan and Korea to promote children's Internet safety. Specifically, there is special attention for the regional trends and initiatives unique to each country in this regard. The findings of this study indicate that the measures to promote Internet safety are reflective of, and shaped by, each country's particular socio-cultural characteristics.

We would like to thank UCSIA for their significant support in realising this book and all colleagues who contributed to the conference and this publication. Our special thanks go to Inke Du Ry for her language review and assistance in the editing process. We hope that this publication helps the reader in assessing the complex reality in which youngsters use ICT to get in touch with each other and consume or produce information online.

Michel Walrave, Wannes Heirman, Sara Mels,
Christiane Timmerman and Heidi Vandebosch

eYouth

Between Trust, Concerns and Science

Michel WALRAVE

Young people are increasingly using the Internet, mobile telephony and gaming platforms for a variety of purposes. They offer youngsters tremendous new opportunities in the areas of information, education, social contact and entertainment. Teenagers often use these technologies in the privacy of their bedroom, sometimes referred to as the *bedroom culture* (Bovill & Livingstone, 2001). Yet, the fact that many youngsters are increasingly spending time in their private space connected to the Internet, may also lead them to engage in several activities which pose a risk. Moreover, due to the absence of the parents, these activities may slip under the parental radar.

In this context, researchers, child protection advocates and consumer organisations have been voicing their concerns about the various risks youngsters may encounter online. Moreover, policy makers are under pressure to make the right decisions to shape the conditions that will offer children and teenagers a *safer* Internet experience. Although adolescents, who belong to the generation of *digital natives*, are often depicted as *techno-savvies*, questions arise about their ability to accurately assess online risks, to cope with negative online experiences and to maximise opportunities offered by new technologies. Therefore, the role played by parents and educators in addressing specific issues and empowering young users is receiving growing attention.

Confronted with an increasing academic and societal attention on these issues, UCSIA and MIOS University of Antwerp have created a platform to discuss results from recent research and policy initiatives during a two-day international conference. From the numerous papers submitted, the conference organizing committee has reviewed and retained the papers included in this volume.

With this publication, the editors not only wish to address the risks that are extensively studied, but also the opportunities for young people to engage in online relationships and explore their creativity using ICT. In short, this book aims to provide the reader with further insight into what youngsters really do online, which opportunities they seize and how they cope with specific online risks.

We start this introduction with several important questions that will be addressed throughout this chapter: is it possible to categorise (new) chances and risks encountered by young users on the Internet? Which risks are more or less prevalent and what are their possible consequences? These questions are addressed in this introduction with the purpose of outlining the broader context in which the following chapters can be situated.

Categorising Risks & Opportunities

The way in which children and adolescents balance between opportunities and risks when engaging in online activities has increasingly been the focus of attention of academic researchers in recent years. Several typologies of online opportunities and risks have been constructed, giving a clear overview of the issues at stake. Yet, how risks are defined rarely goes beyond the way they are perceived by adults. As Lievrouw and Livingstone (2006) rightly argued: what adults sometimes regard as a risk (meeting strangers online, for instance) is often perceived as an opportunity by children and adolescents (making new friends).

Nevertheless, to depict the various opportunities and risks addressed in this volume, we will first invoke the typology developed in the framework of the EU Kids Online network (Livingstone & Haddon, 2009; Livingstone, Haddon, Görzig, & Ólafsson, 2011). The opportunities and risks in this typology are classified in terms of *content* (in which the child is positioned as recipient), *contact* (in which someone else contacts the child and asks him or her to participate, whether unwillingly or unknowingly) and *conduct* (where the child is an actor).

Another typology, developed in the framework of the Youth Protection Roundtable, is closely related to the first typology. One of the differences is the use of a bi-dimensional (contact/content) rather than a three-dimensional model (content/contact/conduct). This typology differentiates content- from contact-related risks. Some risks may result from the Internet user's own conduct, while others may result from other users' conduct. This framework grasps the complexity of online (risky) situations, where youngsters can be targets but also behave as perpetrators of specific contact-related risks and where they are not only recipients of possibly harmful content but also contribute to it. On the side of opportunities, young people are not merely consumers but also become *prosumers*, actively involved not only in developing online relations but also online content.

By merging the risk/opportunities framework (Livingstone & Haddon, 2009) and the matrix of risks and threats (YPRT, 2009), the following model can be elaborated.

Figure 1: Classification of online opportunities and risks for children and adolescents based on Livingstone & Haddon (2009) and YPRT (2009)

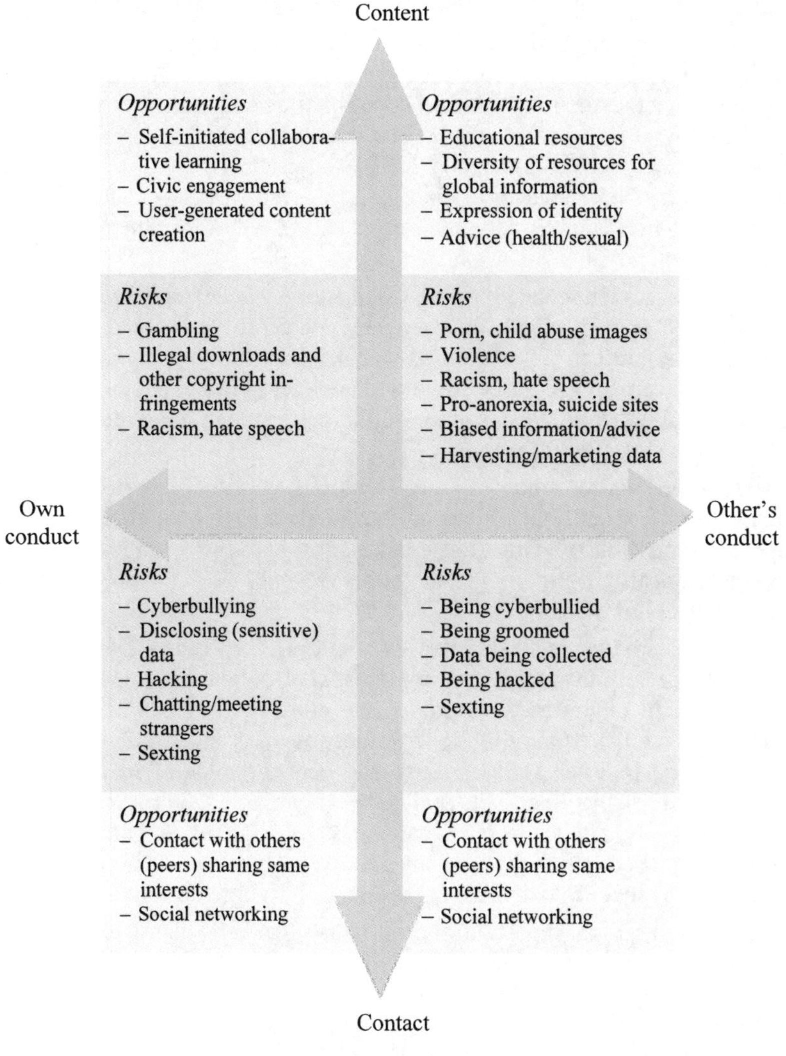

As far as content-related opportunities and risks are concerned, young Internet users can be recipients of educational, entertaining content, but also harmful ones such as pornography, images of violence and racist or other hate speech that is distributed online. They can also be the target of specific marketing techniques, as they are sometimes exposed to advertising or sponsoring of products and services that are not age-appropriate. Other Internet users' conduct can lead to biased information or advice on personal issues. On sensitive topics like anorexia and suicide, young Internet users can consult websites of health-related professional organizations or be misled by pro-anorexia or other websites.

Yet, youngsters can also take initiatives and seize opportunities by using online educational (collaborative) tools, to engage in civic actions (like e-protest, e-petitions, online donations) and to elaborate their own online content (blogging, using social media like Flickr and YouTube). Nevertheless, their conduct can lead to specific risks, like spreading racist or other hate speech online, using gambling sites or engaging in piracy-activities such as downloading music or videos or other copyright infringements.

Next to content-related risks, Livingstone (2010) specifies contact risks as the way in which children are invited to participate in online interactions that can possibly lead to harm. On the opportunities' side, youngsters can engage, or be invited to engage, in (new) social relations, using social network sites or other online platforms to meet other people sharing the same interests.

However, risky online contact situations can not only result from other users' conduct but also from the young user's own conduct. This opens the possibility of making a distinction between youngsters as the target of specific online aggressions on the one hand and their role as perpetrator on the other.

Examples of contact-related risks resulting from other (adult) users' conduct are online sexual grooming, sexual solicitations and harassment. Yet, in other risky situations, the child or teen can be an actor him- or herself, perpetrating specific acts such as cyberbullying. Even more, young Internet users can also augment the odds of specific risks by disclosing (sensitive) personal data, engaging in online chats with older, possibly adult, users. Finally, they can send or (unwillingly) receive sexual explicit material by mobile phone (i.e. sexting), which then leads to specific risks like harassment, bullying etc.

Although online opportunities and risks have been classified, and increasing research efforts are invested in studying the prevalence and possible harm resulting from those risks, it still is a methodologically challenging and ethically sensitive matter to conduct this research especially with children and adolescents (Livingstone *et al.*, 2011). Nonetheless, recent research has been conducted providing insight into online risks that teens are confronted with.

In what follows, we outline the current academic discussions on both opportunities and risks. For each category of opportunities and risks, we will elaborate on one or several topical issues.

Content-related Opportunities & Risks

With regard to online *content* young people are mainly seen as recipients or consumers. They are able to consult educational and news resources that can be useful for performing their school tasks. The importance of the Internet as an educational resource is stressed by the EU Kids Online findings, which state that the use of Internet for school work is a top online activity among European youngsters (85% of the 9- to 11-year-olds) (Livingstone *et al.*, 2011). Although this offers a broad range of possibilities to explore online sources and also to develop a critical attitude towards online information, some teachers and parents doubt if young people are able to critically examine and integrate sources. Moreover, more than eight out of ten 12-to 18-year-olds admit having used online information in their school tasks without referring to the genuine sources (Walrave, Lenaerts, & De Moor, 2009). Therefore, the enhancement of digital competence needs to also encompass the ability to critically examine sources and to refer to them. *Digital literacy* has a multifaceted nature including not only the possibility of selecting the right application for a purpose and of using this application, but also the chance to:

> […] analyze, evaluate […] and integrate digital resources, construct new knowledge, create media expressions and communicate with others, in the context of specific life situations, in order to enable constructive social action and finally, to be able to reflect upon this process (Martin, 2005, pp. 135-136).

Youngsters too recognise the need to practice their information searches and processing more critically. They sometimes find it difficult to decide which websites to search for, some get sidetracked while exploring sources online and admit the temptation of plagiarism (Strom, Strom, Wing, & Beckert, 2010). Moreover, when surveying eight digital skills (among 11-16 year old Europeans), respondents declared that on average they are able to perform four of the skills asked for. Although some differences were found between teenagers and younger children, between girls and boys, the latter claiming to have more skills, some skills (especially e-safety) are not yet widely known or practiced (Livingstone *et al.*, 2011).

Therefore some researchers are sceptical about the over-enthusiastic use of the term *digital natives* and of labelling current children, teens and young adults, born roughly after 1980, as digital experts whereas a majority of their parents are labelled as *digital immigrants* (Prensky,

2001). While the young generation is perceived by some researchers as more skilful in using digital equipment than their educators, others, by contrast, argue that young people are not necessarily experts, especially when using the Internet for educational purposes (Bennett, Maton, & Kervin, 2008; Kennedy, Krause, Judd, Churchward, & Gray, 2006; Levin & Arafeh, 2002; Li & Ranieri, 2010).

In fact, there is a need for heightening the level of teenagers' critical attitude towards resources they use for school work, as well as towards online information that can inspire them in their quest to solve personal questions, doubts and problems. As adolescence is a period of self-exploration and identity construction, an increasing number of online sources and services offer them support. Not only safer Internet help-lines and awareness centres, but also youth advisory organisations specialised in sexual or mental health are taking steps to approach their young public through ICT, by offering online resources and mediation for instance. Some organisations are present in social network sites, others offer counselling using chat or texting. These new channels for advice lower the threshold for youngsters who are in need of advice in order to cope with the problems they face. Besides the accessibility of services, it also enhances therapeutic relationships between youngsters and their advisors/counsellors (Furber *et al.*, 2011).

Young people also increasingly express their identity and explore their creativity online. In this way they themselves become *producers* of online content. The use of social media, which enables them to present themselves and to share pictures and videos, is especially important to young people. In Europe as well as in the United States, the majority of teenagers use social network sites (Livingstone *et al.*, 2011; Lenhart, Purcell, Smith, & Zickuhr, 2010). These online platforms are of great importance to the daily lives of teenagers.

At the same time, on the flipside we call risk, children and teens are seen as targets of advertising through social media and other websites. Sometimes they are confronted with advertisements for certain products and services that are not age-appropriate. A growing body of empirical research analyses the way in which children and teens process commercial or persuasive messages. Yet, this academic research is still mostly concentrated on mass media (television) advertising (e.g. Brucks, Armstrong, & Goldberg, 1988; Buijzen, 2007; Livingstone & Helsper, 2006; Rozendaal, Buijzen, & Valkenburg, 2010). Nevertheless, there is increasing scientific interest for teenagers' response to online marketing messages and how they deal with the disclosure of their personal data for online marketing purposes (Ali, Blades, Oates, & Blumberg, 2009; Chung & Grimes, 2005; Nairn & Dew, 2007; Neeley, 2007; Youn, 2005; Youn, 2008).

In fact, young people's multi-faceted Internet usage and the steady increase in children's and teenagers' discretionary income during the past decades makes them a segment in the commercial market which is difficult to ignore (Montgomery, 2001, Yarrow & O'Donnell, 2009).

Minors do not only have their own budget to spend, they also significantly influence the decision-making process in purchasing household products, something referred to as *kidsfluence* (Sutherland & Thompson, 2001). Finally, research has pointed out that consumer patterns established during childhood years are often consolidated during adulthood. Therefore, it is important for companies to build up long-term customer relationships with children and to develop brand loyalty (Furnham, 2000).

In sum, children and teens are important for marketers in three ways: as (1) actual customers or users of products and services, as (2) able to influence their parents and peers and finally (3) for establishing medium and long term customer relationships. In this regard, companies will try to consolidate their position in the *top of mind* brands of these consumers still growing up.

Following on the need to target teenagers as an influential retailing demographic, there has been a steep increase in commercially fuelled websites dedicated to children (Neeley, 2007). For approximately two out of three children's websites, advertising is the backbone of revenue (Montgomery, 2001). At the same time, the share of non-commercial child-oriented websites has decreased seriously (Neuborne, 2001).

The growth in child-targeting websites employing Internet advertising has, however, generated concern among children's advocates (e.g. Austin & Reed, 1999; Neeley, 2007; Wilcox, *et al.*, 2004). They argue that the commercialisation of young people's online environment has several questionable aspects, like the *hybridisation* of marketing, blending commercial or persuasive messages with entertainment (e.g. advergames). Moreover, the fairness of processing minors' personal data for direct marketing purposes can be questioned.

These concerns are also linked with observations made about the ability of children and teens to make a distinction between medium content and commercial messages and the ability to recognise the persuasive intent of advertising. Both abilities have been perceived as *a cognitive shield* that at least partially protects minors from undesirable advertising effects.

Based on Piaget's theory of cognitive development (Piaget & de Saussure, 1956), several stages in the consumer socialisation of children and teens have been discerned (John, 1999). In the context of TV ads, children between the ages of four and five were found to be able to distinguish commercials from other programmes. Yet, they may not be able to differentiate the intentions of programme types, and merely conceive of advertising as another form of entertainment. Around the age of eight, children are able to perceive the persuasive intentions of advertising. At a later point of time in childhood, more specifically around the age of twelve, teenagers develop a critical attitude towards persuasive (commercial) messages (Robertson & Rossiter, 1974; Rozendaal *et al.*, 2010; Valkenburg & Cantor, 2001). However, recent research doubts if 12-year-olds have a level of understanding of advertising that is *comparable to adults*. It was found that, although a majority of 12-year-olds understand the persuasive and selling intent of advertising, they still have not acquired an *adult level of understanding*. Yet, an adult-like level of *recognition* of advertising was attained from the age of ten or eleven (Rozendaal *et al.*, 2010).

So far, the majority of studies have focused on television advertising. Research into online forms of advertising is of the utmost importance, as some types of online marketing are not straightforward, mingling persuasion and entertainment. Moreover, in contrast with TV commercials, some forms of online marketing do not use (audio)visual *cues* that enable (young) consumers to make the distinction between commercial messages and other forms of content.

Although research on online advertising processing by children is scarce, a recent study reported that 6-year-olds were only able to recognise a quarter of the advertisements compared to half of 8-year-olds and 75% of the 10-12-year-olds (Ali *et al.*, 2009).

Moreover, due to their age, children and teens do not yet possess fully developed consumer skills. These skills are not only important to discern advertising messages and to develop a critical attitude towards commercial persuasion, but they may also help to assess the consequences of personal data requests in (commercial) websites. These requests might make youngsters particularly vulnerable to some privacy-related risks on the Internet such as the profiling of consumers for marketing purposes (Moscardelli & Divine, 2007).

Contact-related Opportunities & Risks

Whereas online risk-related typologies focus on contact risks where young Internet users are merely seen as a target of unwanted (adult initiated) contact, teens can have an active role in contact-related issues too. It is best if the importance of online communication platforms for contact opportunities in general is underlined first.

As far as online contact is concerned, several studies have found that platforms like Instant Messaging programs (IM), chat boxes and Social

Network Sites (SNS) mostly offer an online extension of communication with existing (offline) friends and acquaintances. The most important motives to create an account are to stay in touch with friends and acquaintances, and especially to communicate with school friends after class hours (Van Cleemput, 2010; Vandoninck, d'Haenens, De Cock, & Donoso, 2010; Walrave & Heirman, 2011b). However, a majority of teenagers also engage in new relationships with persons they do not know offline (D'hanens, Courtois, Mechant, & Paulussen, 2010).

Moreover, when creating an account and updating profile information, users can choose which personal information to fill in and manage who will have access to these data, a broad group (e.g. all SNS users) or a narrow group of other users (e.g. their friends and possibly, friends of friends). The information that is shared in a social network site and revealed to others can be defined as *self-disclosure* (Krasnova, Spiekermann, Koroleva, & Hildebrand, 2010). The access to personal data can be managed using *privacy settings*. Yet, half of the younger (sometimes under-age) users and a quarter of older teenage users admit they are not able to change their privacy settings (Livingstone, Ólafsson, & Staksrud, 2011).

Based on social exchange theory, individuals will build interpersonal relationships based on the personal (subjective) evaluation of costs and benefits (Homans, 1958). In social network sites, but also on other websites processing personal data, users will assess whether the benefits of data disclosure exceed the possible risks, something that is sometimes called the *privacy calculus*. Findings of previous research into the privacy concerns of adolescents and adults have shown that next to risk appraisal, the perceived benefits of information disclosure are also important ingredients in the decision-making process of disclosing personal information (Culnan & Armstrong, 1999; Dinev & Hart, 2006; Liebermann & Stashevsky, 2002; Walrave & Heirman, 2011a; Youn, 2005).

Previous studies on the benefits of online data disclosure have revealed that the *intrinsic pleasure* of disclosing information and of presenting oneself online helps to explain the users' decision to self-disclose online. Moreover, the opportunity and convenience offered to keep up with existing contacts as well as to engage in new relationships and to send updates to a group of friends, can motivate users to disclose personal data (Boyd & Ellison, 2007; Ellison, Heino, & Gibbs, 2006; Krasnova *et al.*, 2010; Sledgianowski & Kulviwat, 2008). More particularly it was observed that an individual's disposition to disclose personal information affects positively the use of the social network site, which in turn influences the user's disposition to self-disclose. These effects are moderated by gained friendships and social capital that function as social rewards (Trepte & Reinecke, 2011).

Besides, the more users disclose personal information, the more they facilitate other users in finding common ground for getting in touch (Lampe, Ellison, & Steinfield, 2007). Users can also, in some way, manage their *image* by deciding which information, in words, pictures and group memberships, is posted. The asynchronous communication and the emphasis on verbal communication offers opportunities for *impression management*, as users can take the time to decide how to formulate a reaction and which picture to post (Boyd & Ellison, 2007; Ellison *et al.*, 2006).

With regard to possible privacy concerns of SNS users, contrasting results have been found. On the one hand, a significant negative correlation was observed between the level of privacy concern and self-disclosure, revealing that users who are more concerned about possible privacy threats decide to disclose less personal data and restrict the access to the personal data (Dwyer, Hiltz, & Passerini, 2007; Krasnova *et al.*, 2010; Utz & Krämer, 2009; Walrave & Heirman, 2011b). On the other hand, some authors found a so-called *privacy paradox*, namely a discrepancy between the online privacy concerns claimed and the amount of data actually disclosed (e.g. Acquisity & Gross, 2006). Yet, it was also found that the level of trust in SNS providers and in other SNS members alleviated privacy risk perception, as it correlates with perceived control in managing access to their data and removing improper comments (e.g. Krasnova *et al.*, 2010).

Although social media offer great opportunities for online contact, a lot of media and policy attention is given to (adult) users who have abusive and criminal intentions and cross the online path of minor users. It was observed that young people who created an online profile were more likely to be contacted online by unknown people (peers or adults). However they were not more likely to experience contacts which they perceived as *scary* or *uncomfortable* (Lenhart & Madden, 2007). Other researchers stress that it is not specifically the widespread habit of young people to disclose personal information in social network sites that increases risks; rather, the odds of aggressive sexual solicitations are augmented by the interaction of youngsters with unknown people (in chat rooms, for instance), sending them personal information or talking online with them about sex (Mitchell, Finkelhor, & Wolak, 2007; Ybarra & Mitchell, 2008).

Parents and policy makers are especially concerned about minors who meet strangers online and subsequently offline. One phenomenon has become the focal point of their concern, namely (online) *sexual grooming*: "The process where a child is befriended by an adult in an attempt to gain the child's confidence and trust in order to lower the child's inhibitions in preparation of sexual exploitation and/or abuse online and/or offline" (Eneman, Gillespie, & Bernd, 2010, p. 2).

Although this process of grooming is not a new phenomenon, technology provides abusers with possibilities for browsing the net looking for potential victims, accessing personal (contact) details of possible targets and concealing or falsifying their real identity to establish contact. Yet, some research has found that in the majority of examined cases, victims were aware of the adult age of the person they were speaking to on the Internet. Moreover, offenders were rarely found to be deceiving their victims about their age and their sexual interests. Besides, older teens (13- to 17-year-olds) were found to be more victimised than younger teens and children (Wolak, Finkelhor, & Mitchell, 2004). One of the possible explanations is that, as teens grow older and gain more online experience, they engage in more complex and more interactive Internet use. In fact, older teens are found to be more likely to meet their new online acquaintances in the offline world (Livingstone & Helsper, 2007). Recent data have confirmed that, while less than one third of European children and teens have made contacts online with people they previously did not know offline, less than one in ten have met face-to-face. Significantly older teenagers had agreed with such an offline meeting (Livingstone *et al.*, 2011).

However, the fact that teenagers get acquainted online with new people does not necessarily imply online and offline risks or (physical) harm for them. New online contacts can be peers and indirect friends (friends-of-friends). These people are in some way already indirectly connected to the child's or adolescent's social circle. That is why some studies specifically examined sexually explicit requests from strangers, or acquaintances targeting adolescents online. Several studies have found that on average one in ten adolescents had this experience (Baumgartner, Valkenburg, & Peter, 2010; Mitchell, Wolak, & Finkelhor, 2008; Walrave *et al.*, 2009; Wolak, Finkelhor, Mitchell, & Ybarra, 2008). Significantly more female adolescents than male report having been sexually solicited on the Internet unwillingly (Baumgartner *et al.*, 2010). Although the incidents are often confined to the Internet and reported as relatively mild, a small proportion of youth received aggressive solicitations that could spill over into their offline lives as the solicitor attempted to, or effectively managed to, make offline contact with them (Mitchell *et al.*, 2008).

Next to unwanted solicitations by adults, minors can be the targets, or even the perpetrators, of peer aggression. One of the conduct-related risks that has been scrutinised in recent years is cyberbullying.

Cyberbullying, as an emerging online form of traditional bullying, can be defined as "the use of information and communication technologies to support deliberate, repeated, and hostile behaviour by an individual or group, that is intended to harm others" (Belsey, 2008). This definition includes three characteristics that are summarised in other definitions too, namely the use of ICT, the intention to harm and the repetition of the perpetrated acts (e.g. Juvonen & Gross, 2008).

Yet, some authors who have defined traditional (offline) and cyberbullying also include a possible *power imbalance* between perpetrator and victim, where the latter cannot easily defend him- or herself (e.g. Olweus, 1999; Smith *et al.*, 2008); whereas others enumerate a list of electronic media to perpetrate these acts (e.g. Kowalski & Limber, 2007; Li, 2007). There is still disagreement about how to define and characterise cyberbullying. Moreover, the use of varying definitions and concepts such as online harassment, digital bullying and electronic bullying, concrete measurements (i.e. asking a person *directly* whether he or she has had experience with *cyberbullying* or asking a person whether he or she has been *involved in a range of online activities* that the researcher considers as forms of cyberbullying, in a short or more extended time period and so on), the ages of the respondents, the precise methods used (i.e. online, class or face-to-face surveys) in different studies, have led to huge differences in prevalence rates (Heirman & Walrave, 2008; Tokunaga, 2010; Vandebosch & Van Cleemput, 2008).

Cyberbullying has been found to be a multifaceted phenomenon. It can take different direct (in-the-face) and indirect (behind-the-back) forms. Direct forms include attacking the ICT-property of the victim (also called *physical* cyberbullying, by sending viruses, compromising files etc.), verbal aggression, like flaming or sending other threatening mails or texts, but also non-verbal forms, such as sending obscene pictures. Finally, victims can be socially bullied by being ignored or excluded from an online group (so called *muting*, *banning* or *defriending*). Regarding indirect forms of cyberbullying, perpetrators can decide to reveal sensitive and intimately disclosed personal information of the victim to large audiences (i.e. *outing*). Moreover, the target of cyberbullying can be deceived by the perpetrator pretending to be someone else (i.e. *masquerade*) or spreading gossip behind the back of the victim (Spitzberg & Hoobler, 2006; Vandebosch & Van Cleemput, 2009; Willard, 2007).

Whereas much research has been conducted about the prevalence of cyberbullying and the characteristics of bullies and their victims, less is known about the medium and long term consequences (e.g. Beran & Li, 2007; Patchin & Hinduja, 2006). Moreover, research has concentrated on the victims and perpetrators, whereas fewer authors have devoted their research efforts to the role of bystanders in cyberbullying. Just as in offline bullying bystanders may have an important role in cyberbullying episodes (Salmivalli, Lagerspetz, Björkqvist, Österman, & Kaukiainen, 1996; Fawzi & Goodwin, 2011). On the one hand they can decide to help the bully by assisting or encouraging him (or as Willard (2007) calls them: *bystanders-who-are-part-of-the-problem*). On the other hand bystanders can decide to ignore the bullying scene, whilst others may engage in defending the victim (and therefore become *bystanders-who-are-part-of-the-solution* (Willard, 2007)).

In addition to the research regarding the different roles in cyberbullying and especially the intertwinement of perpetration and victimization in (cyber)bullying (e.g. Li, 2007; Vandebosch & Van Cleemput, 2008; Ybarra & Mitchell, 2004), some authors have examined the particular features of cyberbullying compared to traditional bullying (e.g. Heirman & Walrave, 2008; Kowalski, Limber, & Agatston, 2008; Mishna, Saini, & Salomon, 2009; Patchin & Hinduja, 2006; Slonje & Smith, 2007). Unlike traditional bullying, cyber victims are confronted with the difficulty of getting away from it, leading to a 24/7-attainability. Moreover, cyberbullying acts are sometimes harder to detect in comparison with traditional bullying in the schoolyard, as they slip under the radar of teachers and parents. Besides, rumours and embarrassing pictures are rapidly spread online or via mobile phone and can sometimes attain a larger audience than offline forms of verbal and non-verbal bullying. For some forms of (indirect) cyberbullying, the perpetrator can be less aware of the impact of his acts, as direct feedback from the victim is lacking. This can lead to a lack of empathy and increasing cruelty in subsequent acts. Bullies can also decide to anonymously perpetrate their acts. In this way they increase the victim's distress by not knowing the identity of the perpetrator. In fact, many cyberbullying victims are not aware of the identity of the perpetrator(s) (Dehue, Bolman, & Vollink, 2008; Vandebosch *et al.*, 2006).

Still, a majority of victims decide not to tell their parents or other trusted peers about ongoing online bullying episodes (e.g. Aricak *et al.*, 2008; Dehue *et al.*, 2008; Li, 2007; Vandebosch *et al.*, 2006). The fear that adults do not care or won't believe their story, can prevent some victims from sharing their negative online experiences with adults. Further, the decision to remain silent about cyberbullying is sometimes caused by the victim's fear that his or her parents would take away computer privileges in response to the acts of cyberbullying (Li, 2007; Mishna *et al.*, 2009). Unfortunately, support and intervention opportunities are seriously restricted if the victim does not say anything about the cyberbullying.

Some Final Thoughts

Having summarised and confronted some of the research dealing with online risks and opportunities, we would like to end with a few reflections on the current scientific and public debate.

Although a growing body of knowledge is available on ICT uses by children and teenagers, more insight is needed into the balance, or possible imbalance, of risks and opportunities. Therefore, future research could try to grasp the double-sided nature of ICT use, as risks and opportunities are often not separated but two faces of the same coin. For instance, while sharing personal information offers possibilities to engage in (new) relationships, it can make youngsters more vulnerable for unwanted contact or (online) bullying. While a lot of valuable content for information and education purposes is only a few clicks away, the same ease of access holds for possibly harmful content. More research efforts are actually needed in order to be able to compare the experiences of youngsters that could lead to possibly harmful or beneficial outcomes in the mid and longer term. Furthermore, researchers have to keep in mind the *process* of adolescence: this emotional and cognitive development has been linked to a higher propensity for risk-taking behaviour (Steinberg, 2007), impelling adolescents to engage in new thrilling experiences, testing their own and other's boundaries and challenging norms set by adults. This insight contextualises the possible limits of the important role adults can play in supporting youngsters in their online explorations.

Therefore, further investigation could focus on the possibilities as well as the limits of the educational role parents and teachers can play in helping youngsters to enrich several facets of their daily life, by seizing opportunities offered by ICT as well as empowering them to cope with online risks. In other words, research on the role of education, but also of policy, should not be restricted to the risks.

Further insight is needed into the way that teachers as well as parents can support youngsters in exploring and augmenting opportunities made possible by ICT use. In sum, it is about how children and teens can be accompanied in steadily climbing the *ladder of opportunities* (Livingstone & Haddon, 2009), moving from basic (search and communication) to more advanced (interactive and creative) Internet uses and skills, ones that offer a broader range of benefits (Livingstone & Helsper, 2007). As physical access to ICT is becoming less of an issue in some parts of the world, research and policy can concentrate on the diversity and complexity of ICT use leading to several levels of benefits.

With respect to risk, the scientific, policy and broader public debate needs to keep in mind which risks, in what circumstances, lead to harm and which can induce situations wherein youngsters can discover themselves and others, and are able to learn to anticipate, detect and cope with risks in the world surrounding them.

References

Ali, M., Blades, M., Oates, C., & Blumber, F. (2009). Young Children's Ability to Recognize Advertisements in Web Page Designs. *British Journal of Developmental Psychology*, *27*(1), 71-83.

Acquisity, A. & Gross, R. (2006). Imagined Communities: Awareness, Information Sharing and Privacy on the Facebook. In G. Danezis & P. Golle (eds.), *Privacy Enhancing Technologies Workshop (PET)*, *2006*, pp. 36-58.

Aricak, T., Siyahhan, S., Uzunhasanoglu, A., Saribeyoglu, S., Cipak, S., Yilmaz, N., *et al.* (2008). Cyberbullying Among Turkish Adolescents. *CyberPsychology & Behavior*, *11*(3), 253-261.

Austin, J. & Reed, M.L. (1999). Targeting Children Online: Internet Advertising Ethics Issues. *Journal of Consumer Marketing*, *16*(6), 590-602.

Baumgartner, S., Valkenburg, P., & Peter, J. (2010). Unwanted Online Sexual Solicitation and Risky Sexual Online Behavior Across The Lifespan. *Journal of Applied Developmental Psychology*, *31*(6), 439-444.

Belsey, B. (2008). *Cyberbullying*. Retrieved December 3, 2010, from http://www.cyberbullying.ca.

Bennett, S., Maton, K., & Kervin, L. (2008). The "Digital Natives" Debate: A Critical Review of the Evidence. *British Journal of Educational Technology*, *39*(5), 775-786.

Beran, T. & Li, Q. (2007). The Relationship Between Cyberbullying and School Bullying. *Journal of Student Wellbeing*, *1*(2), 15-33.

Bovill, M. & Livingstone, S. (2001). Bedroom Culture and the Privatization of Media Use. In S. Livingstone & M. Bovill (eds.), *Children and Their Changing Media Environment: A European Comparative Study* (pp. 179-200). Mahwah, N.J., USA: LEA's communication series, Lawrence Erlbaum Associates.

Boyd, D.M. & Ellison, N.B. (2007). Social Network Sites: Definition, History and Scholarship. *Journal of Computer-Mediated Communication*, *13*(1), article 11. Retrieved December 2, 2010, from jcmc.indiana.edu/vol13/issue1/boyd.ellison.html.

Boyd, D. (2008). Why Youth (Love) Social Network Sites: The Role of Networked Publics in Teenage Social Life. In D. Buckingham (ed.), *Youth, Identity and Digital Media* (pp. 119-142). Cambridge: MIT Press.

Brucks, M., Armstrong, G.M., & Goldberg, M.E. (1988). Children's Use of Cognitive Defenses Against Television Advertising: A Cognitive Response Approach. *Journal of Consumer Research*, 14, 471-482.

Buijzen, M. (2007). Reducing Children's Susceptibility to Commercials: Mechanisms of Factual and Evaluative Advertising Interventions. *Media Psychology*, 9, 411-430.

Chung, G. & Grimes, S.M. (2005). Data Mining the Kids: Surveillance and Market Research Strategies in Children's Online Games. *Canadian Journal of Communication, 30*(4), 527-548. Retrieved December 3, 2010, from www.cjc-online.ca/index.php/journal/article/viewArticle/1525/1653.

Culnan, M.J. & Armstrong, P.K. (1999). Information Privacy Concerns, Procedural Fairness, and Impersonal Trust: An Empirical Investigation. *Organization Science, 10*(1), 104-115.

Dehue, F., Bolman, C., & Vollink, T. (2008). Cyberbullying: Youngsters' Experiences and Parental Perception. *CyberPsychology & Behavior*, 11, 217-223.

D'hanens, K., Courtois, C., Mechant, P., & Paulussen, S. (2010). *Apestaartjaren 3*. Onderzoeksrapport, Retrieved December 2, 2010, from http://www.apestaartjaren.be.

Dinev, T. & Hart, P. (2006). Privacy Concerns and Levels of Information Exchange: An Empirical Investigation of Intended E-services. *E-Services Journal, 4*(3), 25-59.

Dwyer, C., Hiltz, S.R., & Passerini, K. (2007). Trust and Privacy Concern Within Social Networking Sites: A Comparison of Facebook and MySpace. *Proceedings of the Thirteenth Americas Conference on Information Systems*. Keystone: Colorado.

Ellison, N., Heino, R., & Gibbs, J. (2006). Managing Impressions Online: Self-presentation Processes in the Online Dating Environment. *Journal of Computer-Mediated Communication, 11*(2), article 2. Retrieved December 2, 2010, from http://jcmc.indiana.edu/vol11/issue2/ellison.html.

Eneman, M., Gillespie, A.A., & Bernd, C.S. (2010). Technology and Sexual Abuse: A Critical Review of an Internet Grooming Case. *ICIS 2010 Proceedings*, paper 144. Retrieved December 2, 2010, from http://aisel.aisnet.org/icis2010_submissions/144.

Fawzi, N. & Goodwin, B. (2011). Witnesses of the Offense: What Influences the Behavior of Bystanders of Cyberbullying? *Proceedings of the 2011 ICA Conference*. Retrieved May 27, 2011, from http://convention2.allacademic.com/one/ica/ica11/.

Furber, G.V., Crago, A.E., Meehan, K., Sheppard, T.D., Hooper, K., Abbot, D.T., *et al.* (2011). How Adolescents Use SMS (Short Message Service) to Micro-Coordinate Contact with Youth Mental Health Outreach Services. *Journal of Adolescent Health*, 48, 113-115.

Furnham, A. (ed.) (2000). *Children and Advertising*. London: Social Affair Unit.

Gillespie, A. (2002). Child Protection on the Internet: Challenges for Criminal Law. *Child and Family Law Quarterly*, 14, 411-425.

Hasebrink, U., Livingstone, S., Haddon, L., & Ólafsson, K. (2009). *Comparing Children's Online Opportunities and Risks Across Europe: Cross-National Comparisons for EU Kids Online*. London: LSE, EU Kids Online, 2nd ed. Retrieved December 2, 2010, from http://eprints.lse.ac.uk/24368.

Heirman, W. & Walrave, M. (2008). Assessing Concerns and Issues About the Mediation of Technology in Cyberbullying. *Cyberpsychology: Journal of Psychosocial Research on Cyberspace*, *2*(2), article 1. Retrieved December 3, 2010, from http://cyberpsychology.eu.

Homans, G.C. (1958). Social Behavior as Exchange. *American Journal of Sociology*, 63, 597-606.

Juvonen J. & Gross, E.F. (2008). Extending the School Grounds? – Bullying Experiences in Cyberspace. *Journal of School Health*, *78*(9), 496-505.

Kennedy, G., Krause, K.-L., Judd, T., Churchward, A., & Gray, K. (2006). *First Year Students' Experiences with Technology: Are They Really Digital Natives?* Preliminary Report Findings, University of Melbourne. Retrieved December 3, 2010, from http://www.bmu.unimelb.edu.au/research/munatives/natives_report2006.pdf.

Kowalski, R.M. & Limber, S.P. (2007). Electronic Bullying Among Middle School Students. *Journal of Adolescent Health*, 41, s22-s30.

Kowalski, R.M., Limber, S.P., & Agatston, P.W. (2008). *Cyber Bullying: Bullying in the Digital Age*. Malden, MA: Blackwell Publishing Ltd.

Krasnova, H., Spiekermann, S., Koroleva, K., & Hildebrand, Th. (2010). Online Social Networks: Why We Disclose. *Journal of Information Technology*, 25, 109-125.

Lampe, C., Ellison, N., & Steinfield, C. (2007). A Familiar Face(book): Profile Elements as Signals in an Online Social Network. *CHI 2007 Proceedings, Online Representation of Self*, 435-444.

Lenhart, A. & Madden, M. (2007). *Teens, Privacy & Social Networks*. Washington DC: Pew Internet & American Life Project. Retrieved December 2, 2010, from http://www.pewInternet.org/Reports/2007/Teens-Privacy-and-Online-Social-Networks.aspx.

Lenhart, A., Purcell, K., Smith, A., & Zickuhr, K. (2010). *Social Media & Mobile Internet Use Among Teens and Young Adults.* Washington DC: Pew Internet & America Life Project. Retrieved December 2, 2010 from http://www.pewInternet.org/Reports/2010/Social-Media-and-Young-Adults.aspx.

Levin, D. & Arafeh, S. (2002). *The Digital Disconnect: The Widening Gap Between Internet-savvy Students and Their Schools.* Washington DC: Pew Internet & American Life Project. Retrieved December 2, 2010, from http://www.educause.edu/Resources/TheDigitalDisconnectTheWidenin/153789.

Li, Q. (2007). Bullying in the New Playground. *Australasian Journal of Educational Technology*, *23*(4), 435-455.

Li, Y. & Ranieri, M. (2010). Are "Digital Natives" Really Digitally Competent? A Study on Chinese Teenagers. *British Journal of Education Technology*, *41*(6), 1029-1042.

Liebermann, Y. & Stashevsky, S. (2002). Perceived Risks as Barrier to Internet and E-commerce Usage. *Qualitative Market Research: An International Journal*, *5*(4), 291-300.

Lievrouw, L.A. & Livingstone, S. (eds.) (2006). *Handbook of New Media: Social Shaping and Social Consequences*. London: Sage.

Livingstone, S. & Helsper, E.J. (2006). Does Advertising Literacy Mediate the Effects of Advertising on Children? A Critical Examination of Two Linked Research Literatures in Relation to Obesity and Food Choice. *Journal of Communication*, 56, 560-584.

Livingstone, S. & Helsper, E.J. (2007). Taking Risks When Communicating on the Internet: The Role of Offline Social-psychological Factors in Young People's Vulnerability to Online Risks. *Information, Communication & Society*, *10*(5), 619-644.

Livingstone, S. & Haddon, L. (eds.) (2009). *Kids Online. Opportunities and Risks for Children*. Bristol: The Polity Press.

Livingstone, S. (2010). eYouth: (Future) Policy Implications: Risk, Harm and Vulnerability Online. Keynote at *eYouth: Balancing Between Opportunities and Risks*, University of Antwerp. Retrieved December 2, 2010, from http://eprints.lse.ac.uk/27849.

Livingstone, S. & Helsper, E. (2010). Balancing Opportunities and Risks in Teenagers' Use of the Internet. *New Media & Society*, *12*(2), 309-329.

Livingstone, S., Haddon, L., Görzig, A., & Ólafsson, K. (2011). *Risks and Safety on the Internet. The Perspective of European Children.* Full findings. London: LSE, EU Kids Online. Retrieved January 17, 2011, from http://www2.lse.ac.uk/media@lse/research/EUKidsOnline/Home.aspx.

Livingstone, S., Ólafsson, K., & Staksrud, E. (2011). Social Networking, Age and Privacy. London: LSE, EU Kids Online. Retrieved April 20, 2011, from http://www2.lse.ac.uk/media@lse/research/EUKidsOnline/Home.aspx.

Martin, A. (2005). DigEuLit – A European Framework for Digital Literacy: A Progress Report. *Journal of Eliteracy*, 2, 130-136. Retrieved December 2, 2010, from http://www.jelit.org/65/01/JeLit_Paper_31.pdf.

Mishna, F., Saini, M., & Solomon, S. (2009). Ongoing and Online: Children and Youth's Perceptions of Cyber Bullying. *Children and Youth Services Review*, *31*(12), 1222-1228.

Mitchell, K.J., Finkelhor, D., & Wolak, J. (2007). Youth Internet Users at Risk for the Most Serious Online Sexual Solicitations. *American Journal of Preventive Medicine*, 32, 532-537.

Mitchell, K.J., Wolak, J., & Finkelhor, D. (2008). Are Blogs Putting Youth at Risk for Online Sexual Solicitation or Harassment? *Child Abuse & Neglect*, 32, 277-294.

Montgomery, K.C. (2001). Digital Kids: The New On-line Children's Consumer Culture. In D.G. Singer & J.L. Singer (eds.), *Handbook of Children and the Media* (pp. 635-650). Thousand Oaks: Sage.

Moscardelli, D. & Divine, R. (2007). Adolescent's Concern for Privacy When Using the Internet: An Empirical Analysis of Predictors and Relationships with Privacy-Protecting Behaviours. *Family and Consumer Sciences Research Journal*, *35*(3), 232-252.

Nairn, A. & Dew, A. (2007). Pop-ups, Pop-unders, Banners and Buttons: The Ethics of Online Advertising to Primary School Children. *Journal of Direct, Data and Digital Marketing Practice*, 9, 30-46.

Neeley, S.M. (2007). Internet Advertising and Children. In D.W. Shuman & E. Thorson (eds.), *Internet Advertising: Theory and Research* (pp. 343-362). Mahwah, New Jersey: Lawrence Erlbaum Associates.

Neuborne, E. (2001). For Kids on the Web, It's an Ad, Ad, Ad, Ad World. *Business Week*, 3745, 108-109.

Olweus, D. (1999). Sweden. In P.K. Smith, *et al.* (eds.), *The Nature of School Bullying: A Cross-National Perspective* (pp. 7-27). London: Routledge.

Patchin, J. & Hinduja, S. (2006). Bullies Move Beyond the Schoolyard: A Preliminary Look at Cyberbullying. *Youth Violence and Juvenile Justice*, *4*(2), 148-169.

Piaget, J. & de Saussure, R. (1956). *Le problème des stades en psychologie de l'enfant: symposium de l'Association de psychologie scientifique de langue française.* Paris: Presses Universitaires de France.

Prensky, M. (2001). Digital Natives, Digital Immigrants. *On the Horizon*, *9*(5), 1-6.

Robertson, T.S. & Rossiter, J.R. (1974). Children and Commercial Persuasion: An Attribution Theory Analysis. *Journal of Consumer Research*, *1*(1), 13-20.

John, D.R. (1999). Consumer Socialization of Children: A Retrospective Look at Twenty-five Years of Research. *Journal of Consumer Research*, 26, 183-213.

Rozendaal, E., Buijzen, M., & Valkenburg, P. (2010). Comparing Children's and Adults' Cognitive Advertising Competences in the Netherlands. *Journal of Children and Media*, *4*(1), 77-89.

Salmivalli, C., Lagerspetz, K., Björkqvist, K., Österman, K., & Kaukiainen, A. (1996). Bullying as a Group Process: Participant Roles and Their Relations to Social Status Within the Group. *Aggressive Behavior*, 22, 1-15.

Sledgianowski, D. & Kulviwat, S. (2008). Social Network Sites: Antecedents of User Adoption and Usage. *AMCIS 2008 Proceedings.* Retrieved December 2, 2010, from http://aisel.aisnet.org/amcis2008/83.

Slonje, R. & Smith, P.K. (2007). Cyberbullying: Another Main Type of Bullying? *Scandinavian Journal of Psychology*, 49, 147-154.

Smith, P.K., Mahdavi, J., Carvalho, M., Fisher, S., Russell, S., & Tippett, N. (2008). Cyberbullying: Its Nature and Impact in Secondary School Pupils. *Journal of Child Psychology and Psychiatry*, 49, 376-385.

Spitzberg, B.H. & Hoobler, G. (2006). Cyberstalking and Technologies of Interpersonal Terrorism. *New Media & Society*, *4*(1), 71-92.

Steinberg, L. (2007). Risk Taking in Adolescence. New Perspectives from Brain and Behavioral Science. *Current Directions in Psychological Science*, *16*(2), 55-59.

Strom, P., Strom, R., Wing, C., & Beckert, T. (2010). Adolescent Learning and the Internet. *Education Digest*, *75*(6), 10-16.

Sutherland, A. & Thompson, B. (2001). *Kidfluence: Why Kids Today Mean Business*. McGraw-Hill Ryerson.

Tokunaga, R.S. (2010). Following You Home from School: A Critical Review and Synthesis of Research on Cyberbullying Victimization. *Computers in Human Behavior*, 26, 277-287.

Trepte, S. & Reinecke, L. (2011). Social Network Site Use and Self-Disclosure Online: A Model of Reciprocal Influences. *Proceedings of the 2011 ICA Conference*. Retrieved May 29, 2011, from http://convention2.allacademic.com/one/ica/ica11/.

Utz, S. & Krämer, N. (2009). The Privacy Paradox on Social Network Sites Revisited: The Role of Individual Characteristics and Group Norms. *Cyberpsychology: Journal of Psychosocial Research on Cyberspace*, *3*(2), article 2. Retrieved December 2, 2010, from http://www.cyberpsychology.eu.

Valkenburg, P.M. & Cantor, J. (2001). The Development of a Child into a Consumer. *Journal of Applied Developmental Psychology*, 22, 61-72.

Van Cleemput, K. (2010). "I'll See You on IM, Text or Call You": A Social Network Approach of Adolescents' Use of Communication Media. *Bulletin of Science, Technology and Society*, *30*(2), 75-85.

Vandebosch H., Van Cleemput, K., Mortelmans, D., & Walrave, M. (2006). *Cyberpesten bij jongeren in Vlaanderen: onderzoeksrapport, samenvatting en executive overview*. Brussel: viWTA.

Vandebosch H. & Van Cleemput, K. (2008). Defining Cyberbullying: A Qualitative Research into the Perceptions of Youngsters. *Cyberpsychology and Behavior: The Impact of the Internet, Multimedia and Virtual Reality on Behavior and Society*, *11*(4), 499-503.

Vandebosch H. & Van Cleemput, K. (2009). Cyberbullying Among Youngsters: Profiles of Bullies and Victims. *New Media and Society* *11*(8), 1349-1371.

Vandoninck, S., d'Haenens, L., De Cock, R., & Donoso, V. (2010). *SNS Among Flemish Youth: Contact Risks and Opportunities*. Paper presented at the eYouth Conference, University of Antwerp.

Walrave, M., Lenaerts, S., & De Moor, S. (2009). *Tieners, ouders en ICT: overzicht resultaten survey*. University of Antwerp: MIOS.

Walrave, M. & Heirman, W. (2011a). Disclosing or Protecting? Teenagers' Online Self-disclosure. In S. Gutwirth, Y. Poullet, P. De Hert, & R. Leenes (eds.) *Computers, Privacy and Data Protection: An Element of Choice* (pp. 285-307). Dordrecht: Springer.

Walrave, M. & Heirman, W. (2011b). *Teenagers' Self-disclosure in Social Network Sites*. Research paper, University of Antwerp: MIOS.

Wilcox, B.L., Kunkel, D., Cantor, J., Dowrick, P., Linn, S., & Palmer, E. (2004). *Report of the APA Task Force on Advertising and Children*. Washington DC: American Psychological Institution.

Willard, N. (2007). *Cyberbullying and Cyberthreats: Responding to the Challenge of Online Social Agression, Threaths, and Distress*. Eugene, OR: Center for Safe and Responsible Internet Use.

Wolak, J., Finkelhor, D., & Mitchell, K.J. (2004). Internet-initiated Sex Crimes Against Minors: Implications for Prevention Based on Findings from a National Study. *Journal of Adolescent Health*, 35, 424.e11-424.e20.

Wolak, J., Finkelhor, D., Mitchell, K., & Ybarra, M. (2008). Online "Predators" and Their Victims. *American Psychologist*, *63*(2), 111-128.

Yarrow, K. & O'Donnell, J. (2009). *Gen BuY: How Tweens, Teens and Twenty-Somethings Are Revolutionizing Retail.* San Francisco, CA: Wiley.

Ybarra, M.L. & Mitchell, K.J. (2004). Online Aggressor/Targets, Aggressors and Targets: A Comparison of Associated Youth Characteristics. *Journal of Child Psychology and Psychiatry*, *45*(7), 1308-1316.

Ybarra, M.L. & Mitchell, K. (2008). How Risky are Social Networking Sites? A Comparison of Places Online Where Youth Sexual Solicitation and Harassment Occurs. *Pediatrics*, *121*(2), e350-e357. Retrieved December 2, 2010, from http://pediatrics.aappublications.org/cgi/content/full/peds.2007-0693v1.

YPRT (2009). *Youth Protection Roundtable Toolkit*, EC Safer Internet Programme. Retrieved December 1, 2010, from http://www.yprt.eu.

Youn, S. (2005). Teenagers' Perception of Online Privacy and Coping Behaviors: A Risk-Benefit Appraisal Approach. *Journal of Broadcasting & Electronic Media*, 49, 86-110.

Youn, S. (2008). Parental Influences and Teens' Attitude Toward Online Privacy Protection. *Journal of Consumer Affairs*, 42, 362-388.

Part I

Access

Investigating Digital Gaps

The Media Richness of Flemish Teenagers

A Multi-method Approach to the Appropriation of Media Technologies

Cédric COURTOIS, Peter MECHANT
& Steve PAULUSSEN

Introduction

In the last decades Western households have increasingly appropriated media technologies. For example, recent representative figures on Flanders, the Northern part of Belgium, indicate that 97% of Flemish people own a television set, 92% has a mobile phone, 81% owns a computer and 78% has a an Internet connection (IBBT, 2009). Consequently, teenagers, growing up in these media-rich environments have been confronted with an abundance of media technologies from a very young age. This generation, for whom a world without technology is probably unimaginable, is often referred to as the *digital natives* (Prensky, 2001), *screenagers* (Rushkoff, 1997) or *the net generation* (Oblinger & Oblinger, 2005). These concepts all refer to their supposed technological savviness and abundant access to various kinds of technologies and devices.

Previous research has indicated the possibility of multifarious consequences from this early acquaintance with media technologies. It has highlighted the role of media in identity formation and the attainment of autonomy, both developmental key issues. For one, media consumption furnishes identity formation, as Brown *et al.* state: "they [adolescents] use media and cultural insights provided by them [media] to see both who they might be and how others have constructed or reconstructed themselves" (Brown *et al.*, 1994, p. 814). Also, independently choosing what content to consume and when to consume it, may provide adolescents with a greater sense of autonomy (Livingstone, 2002).

Yet, these issues are not restricted to the consumption of media content alone. Likewise, access to technology and the actual ownership and use of technology play an equally important role. Even more, drawing upon theoretical insights from domestication theory (Silverstone & Haddon, 1996), we are able to argue that media content cannot be uncoupled from media technologies. In fact, both entities are highly interrelated in making up the meaning attributed to one's appropriation of media. Moreover, given the sustained trend of technological convergence, this interrelation becomes increasingly diverse in its composition. For example, video games can be played not only on a PC or a game console, but also on mobile phones and mobile game devices. Not only do these devices differ strongly in object characteristics and meanings, they also differ in the space in which they are mostly used and hence in which they directly influence the experiences derived from media consumption.

For that reason, it is of the utmost importance not only to know how content is consumed, and how much time is dedicated to this consumption, but also to take into account the devices through which consumption takes place and the context in which technologies are used. This research question gains even more importance when we take into account the possible differences that exist among today's supposedly technology savvy teenagers (e.g. Livingstone, 2007a). With the growing number of possible devices to appropriate, it is not unthinkable that discrepancies among teenagers emerge.

Drawing upon a multi-method methodology, we therefore look in this chapter into media usage and media technology appropriation of Flemish teenagers, with a specific focus on the context of technology use. This paper draws upon the results of the Apestaartjaren 3 research project (www.apestaartjaren.be), a joint effort of two Flemish non-profit organisations (Jeugdwerknet VZW and Graffiti Jeugddienst VZW) and the research group for Media & ICT (MICT) at Ghent University (www.mict.be).

The Domestication of Media Technologies

First of all it is important to sketch out the process by means of which technologies are given a meaningful place in everyday life. As they are increasingly used as social devices, information and communication technologies extend beyond their material and functional nature and also obtain symbolic meanings. For such meanings to be formed, a complex and multidimensional process, referred to as *domestication*, needs to take place. According to Silverstone and Haddon (1996), three related dimensions are to be identified in this process: (a) commodification, (b) appropriation and (c) conversion.

In the commodification phase, which is essentially a commercial process, technologies in the form of products enter the public domain. Through marketing, attempts are made to assign meanings of desire to these products, aiming to instigate consumption. When consumers to some extent accept these attributed meanings and acknowledge their relevance, the process of appropriation can commence. In this active process, a technology is objectified, granting it a comfortable physical and discursive space. Also, appraisals of a technology's functions are incorporated into the ecology of everyday life, possibly even affecting its organisation. However, it is important to emphasise that these functions are not necessarily the same as those intended by the technology's creators or vendors. Finally, the conversion dimension refers to consumer attempts to legitimate the appropriation by re-entering the public domain, seeking both status and belonging by drawing upon shared meanings.

Yet, the meanings ascribed to media are twofold: on the one hand, media consist of a material object, while on the other hand they contain messages. In domestication theory, the concept of a double articulation emphasises that media technologies are both objects and conveyers of messages (Hartmann, 2006). This first articulation refers to the material object itself. The second articulation, less definite, could be described as the content that is consumed using these material objects. Individuals ascribe meaning to both articulations. This implies that a bond with both media and its consequences is rooted in an interrelation of media technologies and media texts. However, in communication studies, research has traditionally focussed on media texts, neglecting the (technological) context in which they are consumed.

Ascribing Meaning to Media Content

As argued in the introduction, both identity and autonomy are central developmental issues. Baumeister (1999) succinctly defines identity as *a definition of the self.* One condition for this definition is that it should differentiate between individuals and be continuous over time. Consequently, anything that supports this differentiation and continuity has the potential of being an attribute of identity. In sum, identity comprises at the least one's interpersonal self, a concept of potentiality and one's general values, principles and priorities. Autonomy, for its part, refers to an active, process of agency situated in the physical, psycho-social, cognitive and moral domain in which the individual, embedded in varying contexts (e.g. the family) develops towards independence and personal freedom (Spear & Kulbok, 2004).

As mentioned in the introduction, independent selection of media consumption provides autonomy as well as identity formation. For one, the latter can be accomplished through a process of active negotiation which leads to the development of cognitive schemes consisting of interpretative definitions of self and others (Huntemann & Morgan, 2001). In fact, these schemes are very broad since multiple aspects of identity are given by media content, including – among others – gender role and sexual identity (Feiring, 1999; Gaunlett, 2009), social belonging, occupational aspirations and political values (Arnett, 1995).

Examples of studies indicating the identity-forming and defining properties of various media contents are abundant. For instance, Fisherkeller (2002) illustrated how teenagers adapt models from televised fiction and relate and incorporate these into their own identities. When it comes to music, teenagers believe that musical style is informative about the kind of people who are listening to it in terms of personal characteristics and normative values. Moreover, there is evidence that one's self-concept and self-esteem are related to musical preferences (North & Hargreaves, 1999). In fact, (popular) music does not only contain messages conveyed through lyrics (e.g. social commentary), it also embodies distinctive visual images of lifestyle (Connel & Gibson, 2003).

Furthermore, several instances of identity formation through video games are present. For one, Marty and Stromer-Galley (2007) argue, in the context of the simulation video game *The Sims*, that the game offers a basis for identity development and experimentation with social interaction and norms. In this game, players are meant to create a character, build a house and form and maintain relationships. Moreover, playing games induce both intra- and extra-game interaction. Durkin (2006) stresses the functional role of video gaming in the attainment of autonomy. Firstly, teenage video-gamers are able to differentiate their own media use from that of their parents' and mould it into a peer-related activity. Secondly, in games, players experience the symbolic autonomy of agency: they are encouraged to face challenges and make their own choices.

Nonetheless, one of the most important fields of exploration must be the Internet. Various studies have indicated that the Web serves as an experimental playground in which teenagers can experiment with their selves. In their communication, they can pretend to be someone else, have an opposite gender, etc. (Gross, 2004; Valkenburg, Schouten, & Peter, 2005). In fact, popular social network sites such as Myspace and Facebook serve as ideal platforms for these kinds of explorations. These platforms provide large degrees of freedom to present oneself, not only through text (who I am, what I like) but also through various stylistic components such as a background, colour scheme and media components such as video clips and photos. Also, the self is placed within the larger context of a peer network, reflecting the social interconnections of teenagers (Boyd, 2008; Livingstone, 2008).

The examples sketched in the previous paragraphs are of course no more than the tip of the iceberg and serve just to illustrate the meaningfulness of media content for teenage media consumers. However, as we have previously argued, content constitutes only one side of the coin as it only covers the meanings attached to the *second articulation* of media technologies, namely the messages they convey.

The Medium and the Message

In his much-cited work *Understanding Media*, McLuhan (1964) argued that *The Medium is the Message*. This phrase refers to the symbolic collusion between a message and the medium through which it is conveyed. In this technological and determinist perspective, the medium's properties exercise such an influence that the message gets absorbed. Although this is a very strong and contested claim, Silverstone (1994) partially agrees within the context of a double articulation. He acknowledges that a medium becomes the message in such a way that it contributes to the meaning that is attributed to media experience as a whole.

Yet, the question remains in what sense technological objects become factors in the overall meaning of media consumption (*first articulation*). Let us take as the example of mobile communication by teenagers. Although speech communication was nothing new, mobile phones did influence this form of communication and its meaning. Oksman and Turtiainen (2004) studied the appropriation of mobile communication and found that in various cases, parents got their children a mobile phone because of safety reasons: the thought of being able to instantly contact son or daughter provided a measure of control, security and also trust. However, the teenagers themselves sometimes interpreted this kind of control as a form of interference. At the same time the possibility of independent communication with peers gave a measure of autonomy. Furthermore, teenagers massively embraced text messages, seeing them as an even more personal form of mobile one to one communication. Also, Caronia (2005) found that mobile phones are used by teenagers at *no-when-times* in *no-where-spaces* to communicate with friends. These concepts respectively refer to undefined and meaningless entities of time and space, for example, when someone is simply walking to somewhere or waiting at a bus stop. This illustrates how a technology allows the ascription of meaning to previously meaningless entities. Even social spaces laden with meaning such as home or school are affected by this technology.

Other accounts have focussed on the mobile phone as a fashion statement. Katz and Sugiyama (2006) state that a technology's role is not limited to its functions. It might not even be the most prominent consideration in acquiring it. Surveying young people, they noticed that early mobile phone adopters were more fashion-oriented. They also discovered a relationship between heavy usage and the importance attributed to the device's style over its function.

Contextualising Media Use: A Bedroom Culture

So far, we have discussed the importance and interplay of media technologies and media texts. However, one crucial issue has remained unaddressed, namely the context in which media are consumed. In this paper we will focus on the domestic space. According to Livingstone (2002), since the 1950s two distinct and relevant trends have emerged which characterise the significance of the home in the daily life of teenagers. First of all, family life has become clearly separated from public life. This demarcation between public and private space is symbolised by the front door. Secondly, the bedroom door symbolises a relative separation of a teenagers' private space from both family and public life. As Livingstone argues, both trends have been closely related to the appropriation of mass media in the home.

Research has indicated that teenagers voluntarily seek to spend time alone, something that is positively correlated with short-term well-being and positive emotional states (Larson, 1997). The benefits of spending time alone are actually diverse: it can be seen as a creative space; a time-out enabling critical self-reflection and self-regulation (Buchholz, 1997). Moreover, these spaces have become increasingly filled with media technologies. This has led to the formulation of the concept of a *bedroom culture*, implying that "children and young people spend significant proportions of their leisure time at home with the mass media, increasingly screen media, in their own private space rather than communal or family space" (Bovill & Livingstone, 2001, pp. 180-181). Bovill and Livingstone found that a room's degree of media richness is positively correlated with the time a teenager spends in it. Moreover, media-rich teenagers spend less time with family and more time on their own (Bovill & Livingstone, 2001).

Yet, the question remains what makes up the appeal of these media-furnished bedrooms. Livingstone (2002, 2007b) touches upon several motivations. First, a private bedroom is a convenient storage place for all kinds of private goods, among which media-related objects play an important role. Second, a bedroom provides a possibility to escape: a place where one can autonomously consume media, without interruption. Third, the bedroom as a space also aids the formation of identity: the careful selection and arrangement of personal furniture and equipment, reflecting the bedrooms' owner, requires increasing attention and becomes the subject of territorial claims.

Still, not all teenagers gain access to private media devices. And even if they do, this does not necessarily lead to solitary media use. For example, Pasquier (2001) reported a robust tendency to watch television together, despite the availability of personal devices. This raises the question what kind of technologies are attributed communal meaning and which are not.

Inquiry

In the previous sections we have argued that media consumption is a complex interplay of media technologies (first articulation, the object itself), media content (second articulation, the media texts) and the context in which both are combined. Despite numerous accounts of research into media texts, the supporting media technologies have received rather less attention, even though technological objects also carry meaning in the process of media consumption. In this inquiry, we look into the appropriation of media technologies by Flemish teenagers. In doing so, we sketch out a reliable overview of the technologies teenagers own and the meaning they give to them. Moreover, we take into account the possibility of unequal access to media technologies, especially given the increasing number of possibilities.

Method

The present study consists of a two-phased methodology. For the first phase, which comprises a large-scale quantitative survey, twelve Flemish secondary schools agreed to cooperate. These schools, selected to reflect the diversity in types of education, agreed to allocate class time to let their pupils fill out the computer-based questionnaire. A total of 1725 Flemish secondary school pupils participated (M_{age} = 15.24, SD_{age} = 1.85; 49% Female, 51% Male). After data cleaning, a three-dimensional weighing procedure was applied to correct the sample for gender, grade and education type. As such, the analysed dataset reflects the Flemish secondary school population (Departement Onderwijs, 2009). The questionnaire comprised questions regarding socio-demographics and adoption and usage of the following media (technologies): television, gaming, mobile devices and computers. The survey data was analysed using the SPSS 15 and Latent Gold 4.5 software packages.

In the second research phase, a number of forty face-to-face follow-up interviews were undertaken. The interviewees were recruited from the pool of respondents in phase one, using a maximal variety sampling strategy. These semi-structured interviews dealt with the appropriation of the media technologies surveyed in the research of phase one. Moreover, the technique of photo elicitation was used to facilitate the interviews. The narrative data was analysed using NVivo 8.

Results

Media Technology Ownership and Media Use

To grasp teenage media technology ownership, media use and context of use, a latent class analysis was performed on the data surveyed. This statistical technique allows for the discovery of unobserved subgroups within a given set of categorical variables (Vermunt & Magidson, 2006). Such an approach offers a look into latent structure of technology ownership, media use and context, rather than analysing the data with manifest dependent variables. For that reason, it affords a more detailed and comprehensive view of this complex matter.

In this study, a three-class model yielded an excellent model fit ($L^2(1686) = 1463.19$, $p = 1$, $N\text{par} = 39$, BIC = 21401.95). In the following paragraphs we briefly describe these classes, which significantly differ for age ($F(2, 1722) = 30.88$, $p < .001$, partial ç2 = .04) and gender (÷2(2) = 218.21, $p < .001$).

The first class is made up of teenagers with a *media-rich bedroom* (67% Males, $M_{age} = 15.27$, $SD = 1.88$). This group is characterized by relatively high weekly frequencies of watching television, gaming and using the Internet. Moreover, these teenagers share high probabilities of having these media technologies in their personal bedrooms. We also noticed very high probabilities of having a mobile phone and another mobile device such a mobile game console or an MP3 player.

The second class of teenagers comprised teenagers with a *media-rich environment* (55% Males, $M_{age} = 14.77$, $SD = 1.81$). Although their media attendance and ownership closely resembles the one of the first class, the actual media technologies owned were located somewhere else than the bedroom (e.g. the living room or a study).

The third class consists of teenagers with a *limited media environment* (75% Female, $M_{age} = 15.64$, $SD = 1.74$). In comparison with both previous classes, it is characterised by subtle differences in both media usage and ownership. More specifically, watching television and especially playing video games occur far les frequently. Still, Internet usage differed only marginally. Finally, this third class showed a very low probability of having a game console and slightly lower probabilities of having a mobile device, television or computer.

Teenagers with a…	**Media-rich bedroom**	**Media-rich environment**	**Limited media environment**	Wald	R^2
Cluster Size:	38%	32%	29%		
Indicators probabilities:					
Television frequency per week				29.41***	.03
0-15h TV	0.31	0.43	0.48		
16-20h TV	0.17	0.17	0.17		
20h+ TV	0.52	0.39	0.35		
Gaming frequency per week				50.67***	.30
0-2h gaming	0.27	0.46	0.95		
3-10h gaming	0.26	0.26	0.04		
10h+ gaming	0.47	0.29	0.00		
Internet frequency per week				80.44***	.09
0u-<10h	0.19	0.46	0.39		
10u-<16h	0.18	0.23	0.23		
16-<26h	0.26	0.18	0.21		
26h (+)	0.36	0.13	0.18		
Mobile device ownership+				43.95***	.04
No mobile device	0.03	0.09	0.17		
Mobile device	0.97	0.91	0.83		
Mobile phone ownership				6.23*	.01
No mobile phone	0.01	0.04	0.04		
Mobile phone	0.99	0.96	0.96		
Television location				159.97***	.30
No TV	0.00	0.00	0.02		
TV in bedroom	0.79	0.15	0.36		
TV elsewhere	0.21	0.85	0.62		
Computer location				79.55***	.10
No PC	0.00	0.01	0.04		
PC in bedroom	0.60	0.23	0.44		
PC elsewhere	0.40	0.76	0.52		
Gaming console location				245.58***	.47
No game console	0.05	0.14	0.86		
Game console in bedroom	0.68	0.04	0.08		
Game console elsewhere	0.27	0.82	0.07		

+ MP3/Mobile game devices * $p < .05$, *** $p < .001$

The Appropriation of Television

No less than 99% of the surveyed respondents had at least one television set at home. Teenagers with a *media-rich bedroom* reported a viewing time of roughly half an hour more than both other classes the day before the survey took place. Still, despite having a set of their own, these teenagers spent most of the time watching TV in the living room. Moreover, 44% of these *media-rich bedroom* teenagers reported they most often watched in the company of their family. Also, 60% of both other classes claimed to spend most of the time watching in the presence of their families. Nonetheless, substantial proportions in all three classes reported viewing in solitude.

	Media-rich bedroom	**Media-rich environment**	**Limited media environment**	**Total sample**
*Time spend watching TV "yesterday"****	2h18min	1h48min	1h46min	1h59min
*Mostly watch TV in... (in %)****				
Bedroom	29	2	12	16
Living room	67	94	83	80
Elsewhere	4	5	4	4
*Mostly watch TV... (in %)****				
Alone	49	37	33	41
Together with family	44	60	60	54
Together with friends	7	4	7	6

*** $p < .001$

In the qualitative phase, we further elaborated on these issues. We found that teenagers with a television in the bedroom still prefer the living room for two major reasons. Firstly, in general, the living room is better equipped: a larger screen, better sound, personal video recorders, iDTV-connections, more comfortable seats, etc. Secondly, several respondents explicitly stated that watching television is in some cases more fun to do with family: "I don't like it to be alone in the living room… it's much more fun if your parents are there too, so you can talk. If my dad watches TV, I can watch with him when I can't find anything cool on the Internet" (Nick, 14, M, *media-rich environment*).

On the other hand, several interviewees claimed they prefer to watch television in their own bedroom or are quite eager to have one. The most prominent reason is to avoid conflict and awkward situations. Actually, this indicates that watching television has a communal meaning on the one hand, while on the other hand, it can be used to experience autonomy. Kimberly (13, F, *media-rich environment*), for one, expressed the need for this experience:

> Interviewer: Do you sometimes have arguments about what is being watched on TV?
>
> Kimberly: Yes…
>
> Interviewer: Why does this happen?
>
> Kimberly: Just… because, for example, when it's Thursday, Grey's Anatomy is aired. Recently, I wanted to watch, but she [her mother] didn't want to so I got off to my room. Later, when I returned, we did watch the episode, but… she was kind of angry…

The Appropriation of Computers and the Internet

97% of the respondents have at least one computer at home with 97% of those connected to the Internet. Moreover, 69% of the teenagers with a computer at home have a device they can call their own, 94% of which has an Internet connection. Again it shows that teenagers with a *media-rich bedroom* have a far higher proportion of their own devices, while their own computer is mostly used in the bedroom. Still, despite having a similar rate of computers in the household, teenagers with a *limited media environment* appear to have a higher percentage of own computers than teenagers in a *media-rich environment.* Furthermore, 70% of these computers are used in the bedroom. This indicates that despite non-significant differences with respect to having a computer in the home, its location, property and possibly even its use are quite different.

	Media-rich bedroom	**Media-rich environment**	**Limited media environment**	**Total sample**
Teenagers with a PC at home (in %)	98	97	96	97
Teenagers with own PC (in %)***	83	59	65	70
Most frequent location where own PC is used (in %)***				
Bedroom	73	42	70	64
Living room	20	39	21	25
Study	3	6	5	4
Elsewhere	4	13	5	7

*** $p < .001$

In general, the respondents in the interviews did not attribute much importance to the brand or the type of their computer. However, one recurrent phenomenon is discontent with old and slow devices. Also, teenagers find it very important to have a computer of their own. They do not like it when parents or siblings access their files or monitor their behaviour (e.g. activity on a social network site): this is illustrated by the case of Evi (14, V, *media-rich environment*)

> Interviewer: Can others use your computer?
>
> Evi: Euh, no...
>
> Interviewer: Why not?
>
> Evi: There are pictures on it of me and my friends... I don't think anyone should see those, especially not my parents
>
> Interview: So you are the only one who uses that computer?
>
> Evi: Yes, sometimes my sister, for a game, or something like that... Or my friends, to watch pictures and stuff like that. But they'll never use it without me being there.

An Internet PC's practical functions are quite diverse, comprising entertainment (e.g. multimedia consumption, gaming), the maintenance of social relations (e.g. chat), searching for information, etc. The table below shows that, on the one hand, teenagers with a *media-rich bedroom* engage in multimedia activities more often than the other classes, while, on the other hand, teenagers with a *media-rich environment* or a *limited media environment* report higher rates of instrumental use, such as doing homework or searching information.

	Media-rich bedroom	**Media-rich environment**	**Limited media environment**	**Total sample**
Used computer functions (in %)				
Homework***	56	63	68	61
Search for information	74	77	75	75
Listen to music**	89	82	85	86
Watch videos***	85	76	67	77
Chat*	86	80	81	83

*** $p < .001$

The Appropriation of Video Gaming

Video gaming is a very broad concept, covering a wide range of different devices and game genres. On average, the respondents in our sample roughly spent one hour on gaming the day before the survey was administered. Yet, in contrast to watching television and using the Internet, very large differences emerge when gaming frequency is compared across classes. The same goes for a comparison of game devices used: all types of devices are predominantly used by teenagers with a *media-rich bedroom* or *environment*. Still, as discussed in the previous section, these classes differ concerning the location of their computers. The same is true for fixed game consoles: 71% of the teenagers with a *media-rich bedroom* has a private console in the bedroom, while 66% of *media-rich environment* teens have their console located in the living room. However, a small, yet substantial proportion of teenagers with a *relatively media-poor environment* uses a computer or a mobile phone to play games on a regular basis.

	Media-rich bedroom	**Media-rich environment**	**Limited media environment**	**Total sample**
Time spend gaming "yesterday"***	1h27min	52min	22min	1h05min
Use of gaming devices (in %)				
Computer***	80	79	28	64
Fixed game console***	68	55	9	46
Mobile game console***	60	55	9	43
Mobile phone***	47	40	15	35
Game console location (in %) ***				
Bedroom	71	1	13	36
Living room	18	66	1	34
Somewhere else	9	25	2	14
No game console	3	9	85	16

*** $p < .001$

The diversity in gaming devices is also associated with playing different genres in different places. For example, Adeline (14, F, *limited media environment*) usually plays social games like Guitarhero and Singstar with her friends in her room, fighting games with her father in the living room and single-player games such as Tomb Raider when she is on her own, on her computer. Sander (14, M, *media-rich environment*) likes racing games such as Need for Speed, but also first-person shooters (e.g. Battlefield 1945). He likes to play on his PC as much as on his Playstation game console. He also likes online Flash games to pass time. He would even like to have a Playstation Portable: "It's handy to take along with you, when you need to go somewhere. So when you're in the car, you can play".

The Appropriation of Mobile Devices

As our figures show, mobile devices are widely adopted by the respondents in our sample. A total of 85% owns an MP3 player, logically different according to the different classes. Mobile phones, on the other hand, are even more widespread, without any difference among teenagers. Still, the figures below clearly indicate that the functions used on mobile phones strongly diverge. Unlike others, teenagers with a *media-rich bedroom* use online and multimedia applications more often, to listen to music, to watch video clips and to take pictures.

	Media-rich bedroom	**Media-rich environment**	**Limited media environment**	**Total sample**
MP3 player ownership***	92	82	81	85
Mobile phone ownership	99	97	97	98
Mobile phone functions used "last week" (in %)				
Phone calls***	89	81	88	86
Text messages	96	95	97	96
Take pictures***	74	63	63	68
Listen to music***	74	57	57	64
Watch videos***	39	25	17	28
Surf the Web***	16	9	10	12
Update SNS profile***	15	7	10	11

*** $p < .001$

This multifunctional nature of today's mobile phones appears to be one of the main advantages and attractions of these devices. In fact, teenagers are able to substitute or save on other devices. For instance, Sofie (17, F, *media-rich environment*) was especially focussed on the mobile phone's built-in photo camera's quality so she did not have to buy an additional camera. Kimberly (13, F, *media-rich environment*) uses her mobile phone as extra MP3-player: "I use it to go to school, because I'm afraid to lose my iPod. And… otherwise you have to carry two devices".

Notable is the individual nature of mobile media. Mae (13, F, *limited media environment*) explicitly argues that no one should even touch her mobile phone, not even her parents, because "it violates her privacy", which is a component of autonomy. Also, mobile phones as devices have communicative properties. Ziggy (16, M, *media-rich bedroom*) acknowledges mobile phones as status symbols:

Interviewer: Do you sometimes get remarks (concerning your mobile phone)?

> Ziggy: Euhm, yeah, I do… really unbelievable. The first I bought had a normal, regular shell and at that time, those "sliders" were fashionable. And when I bought one of those, the "folders" got trendy. Recently I bought such a phone, and now everybody has one with a touch screen. Unbelievable, I'm always lagging behind.
>
> Interviewer: Who says you are lagging behind?
>
> Ziggy: I think so myself… and others also say: "Come on, what are you doing with such an outdated cell?"

In fact, the figures below show substantial proportions of teenagers who got another mobile phone because of a new design or features. This is especially true of teenagers with a *media-rich bedroom.*

Reasons to get a new mobile phone (In %):	**Media-rich bedroom**	**Media-rich environment**	**Limited media environment**	**Total sample**
It is broken**	60	53	54	59
New design***	39	25	29	32
New features***	32	15	20	23
Promotion	8	6	5	7
As a gift**	24	16	18	20
Other reasons	9	9	12	10
Still got my first mobile***	6	21	10	12

** $p < .005$, *** $p < .001$

Discussion

The research bears some thought-provoking results. First of all it appears that teenagers are fairly heterogeneous in their appropriation of media technologies. They differ in the technologies they own, the amount of content they consume and the environment in which they do this. Further elaboration of these figures through qualitative research, does, however, contextualise the importance of technologies and the context in which they are used. For instance, although many teenagers have a television of their own, a substantial proportion reverts to the living room to watch. The reasons are manifold, whether social and spatial context (e.g. presence of family, more comfortable seats) or technological characteristics (e.g. screen quality and size, better sound, a set-top box, etc.). Even so, teenagers confirm that their very own television provides the benefit of choosing whatever they like to watch, thus avoiding conflicts with other members of the household.

Another example is found in relation to computers and Internet. Although teenagers in most cases do not really care about a computer's brand, they do want it to be perfectly in order, working at a reasonable pace. Context also matters in this case. As reported in the results, teenagers appreciate privacy, for instance when they are interacting on social network sites. In fact, they feel particularly uncomfortable when someone (e.g. a parent) is watching. Teenagers also articulate a desire to have a computer of their own, and when they have one, they regard it as exclusively personal. Just as television, computers can provide a symbolic function of autonomy.

The strongest proprietary claims are made for mobile devices. Analogous to computers, they contain personal information (contacts, photos, etc.). In fact, recent types of mobile phone resemble pocket media centres, not only allowing for communication, but also for media consumption. These personal devices also carry additional meaning as status symbols. If someone does not own a trendy mobile phone, it is commented upon by peers. In contrast to television and computer, mobile phones convey a strong sense of identity. Brand and appearance seem to be very important as they function as a means to differentiate oneself from others. This occurrence is unmistakably related to, and amplified by, the variety of spaces in which such a device is used (not only at home, but also in public spaces). In case of a television and computer, mostly located in the home, this is far less evident.

In this chapter we have demonstrated the large variety and also the differences in the appropriation of media technologies. Several instances explicitly point to the importance of a clear overview of the technologies teenagers use and the context in which they are used. In fact, devices differ in their ascribed meanings: some are shared while others are strictly personal. Some are attributed additional meaning (as a status symbol), while for others only functional criteria have importance. A consistent finding in our research is that all media technologies carry a meaning of autonomy through allowing independent choice as to what content to consume, what files to store and with whom to communicate. Moreover, mobile devices also contribute to identity: teenagers are able to differentiate themselves from others by displaying the decisions they have made in their consumption. These instances confirm the importance of devoting attention to media technologies.

Given growing technological convergence, most of these devices share similar rudimentary functions. For example, televisions, computers, gaming devices and some mobile phones and MP3 players allow the consumption of audiovisual content such as a soap opera. Consequently, the very same soap can be viewed on any device. It is superfluous to add that the properties of these devices and the contexts in which they are used play a role in how media are received. By focussing only on media texts, only one side of the coin is taken into account.

As our figures show the diversity in the appropriation of media technologies, it becomes clear that the first articulation of media technologies, especially in the case of teenagers, cannot be ignored. In this paper we have illustrated their meanings of autonomy and identity. In conclusion, in order to grasp the complexity of teenage media use, we would like to express the need for further media research to incorporate this understanding.

References

Arnett, J.J. (1995). Adolescents' Uses of Media for Self-Socialization. *Journal of Youth and Adolescence, 24*(5), 519-533.

Baumeister, R.F. (1999). Self-concept, Self-esteem, and Identity. In V.J. Derlega, B.A. Winstead, & W.H. Jones (eds.), *Personality: Contemporary Theory and Research* (pp. 339-375). Chicago: Nelson-Hall Publishers.

Bovill, M. & Livingstone, S. (2001). Bedroom Culture and the Privatization of Media Use. In M. Bovill & S. Livingstone (eds.), *Children and Their Changing Media Environment: A European Comparative Study* (pp. 179-200). Mahwah: Lawrence Erlbaum Associates.

Boyd, D. (2008). Why Youth (Heart) Social Network Sites: The Role of Networked Publics in Teenage Social Life. In D. Buckingham (ed.), *Youth, Identity, and Digital Media* (pp. 119-142). Cambridge: MIT Press.

Brown, J.D., Dykers, C.R., Steele, J.R., & White, A.B. (1994). Teenage Room Culture: Where Media and Identities Intersect. *Communication Research, 21*(6), 813-827.

Schaler Buchholz, E. (1997). *The Call of Solitude: Alonetime in a World of Attachment.* New York: Simon and Schuster.

Caronia, L. (2005). Mobile Culture: An Ethnography of Cellular Phone Uses in Teenagers' Everyday Life. *Convergence, 11*(3), 96-103.

Connel, J. & Gibson, C. (2003). *Sound Tracks: Popular Music, Identity, and Place.* London: Routledge.

Departement Onderwijs. (2009). *Statistisch jaarboek van het Vlaams Onderwijs 2008-2009.* Retrieved March 1, 2010, from http://www.ond.vlaanderen.be/onderwijsstatistieken.

Durkin, K. (2006). Game Playing and Adolescent Development. In P. Vorderer & J. Bryant (eds.), *Playing Video Games: Motives, Responses, and Consequences* (pp. 491-506). Mahwah: Lawrence Erlbaum Associates.

Feiring, C. (1999). Gender Identity and the Development of Romantic Relationships in Adolescence. In W. Furman, B.B. Brown, & C. Feiring (eds.), *The*

Development of Romantic Relationships in Adolescence (pp. 211-232). Cambridge: Cambridge University Press.

Fisherkeller, J. (2002). *Growing Up with Television: Everyday Learning Among Adolescents.* Philadelphia: Temple University Press.

Gauntlett. D. (2009). *Media, Gender and Identity. An Introduction.* London: Routledge.

Gross, E.F. (2004). Adolescent Internet Use: What We Expect, What Teens Report. *Applied Developmental Psychology*, 25, 633-649.

Hartmann, M. (2006). *The Triple Articulation of ICTs. Media as Technological Objects, Symbolic Environments and Individual Texts Domestication of Media and Technology.* Berkshire: Open University Press.

Huntemann, N. & Morgan, M. (2001). Mass Media and Identity Development. In D.G. Singer & J.L. Singer (eds.), *Handbook of Children and the Media* (pp. 309-322). Thousand Oaks: Sage.

IBBT. (2009). *Digimeter Wave 1: IBBT – iLab.o.*

Katz, J.E. & Sugiyama, S. (2006). Mobile Phones as Fashion Statements: Evidence from Student Surveys in the US and Japan. *New Media & Society*, *8*(2), 321-337.

Larson, R.W. (1997). The Emergence of Solitude as a Constructive Domain of Experience in Early Adolescence. *Child Development*, *68*(1), 80-93.

Livingstone, S. (2002). *Young People and New Media.* London: Sage.

Livingstone, S. (2007a). Gradations in Digital Inclusion: Children, Young People and the Digital Divide. *New Media & Society*, *9*(4), 671-696.

Livingstone, S. (2007b). From Family Television to Bedroom Culture: Young People's Media at Home. In E. Devereux (ed.), *Media Studies: Key Issues and Debates* (pp. 302-321). London: Sage.

Livingstone, S. (2008). Taking Risky Opportunities in Youthful Content Creation: Teenagers' Use of Social Networking Sites for Intimacy, Privacy and Self-expression. *New Media & Society*, *10*(3), 393-411.

Martey, R.M. & Stromer-Galley, J. (2007). The Digital Dollhouse: Context and Social Norms in The Sims Online. *Games and Culture*, *2*(4), 314-334.

McLuhan, M. (1964). *Understanding Media: The Extensions of Man.* New York: McGraw Hill.

North, A.C. & Hargreaves, D.J. (1999). Music and Adolescent Identity. *Music Education Research*, *1*(1), 75-92.

Oblinger, D.G. & Oblinger, J.L. (2005). *Educating the Net Generation.* Online e-book. Educause. Retrieved March 1, 2010, from http://www.educause.edu/educatingthenetgen/.

Oksman, V. & Turtiainen, J. (2004). Mobile Communication as a Social Stage. *New Media & Society*, *6*(3), 319-339.

Pasquier, D. (2001). Media at Home: Domestic Interactions and Regulation. In S. Livingstone & M. Bolvill (eds.), *Children and Their Changing Media*

Environment: A European Comparative Study (pp. 161-178). Mahwah: Lawrence Erlbaum Associates.

Prensky, M. (2001). Digital Natives, Digital Immigrants, Part II: Do They Really Think Differently? *On the Horizon*, *9*(6), 1-9.

Rushkoff, D. (1997). *Ecstasy Club.* San Francisco: Harper San Francisco.

Silverstone, R. (1994). *Television and Everyday Life*. London: Routledge.

Silverstone, R. & Haddon, L. (1996). Design and the Domestication of Information and Communication Technologies: Technical Change and Everyday Life. In R. Mansell & R. Silverstone (eds.), *Communication by Design* (pp. 44-74). Oxford: Oxford University Press.

Spear, H.J. & Kulbok, P. (2004). Autonomy and Adolescence: A Concept Analysis, *Public Health Nursing*, *21*(2), 144-152.

Valkenburg, P., Schouten, A., & Peter, J. (2005). Adolescents' Identity Experiments on the Internet. *New Media & Society*, *7*(3), 383-402.

Vermunt, J.K. & Magidson, J. (2006). *Latent Class Analysis.* Retrieved February 1, 2010, from http://www.statisticalinnovations.com/articles/Latclass.pdf.

Digital Inequalities Among the Digital Native

The End of a Myth?

Périne BROTCORNE

Introduction

Since digital technologies are widespread across all domains of daily life, increasing academic and policy attention focuses on the so-called *digital divide*, which refers to inequalities within and across societies between those who have access to and use ICT and those who do not (Brotcorne & Valenduc, 2008; Di Maggio *et al.*, 2004; Hargittai, 2002; Selwyn, 2005, 2006; Selwyn & Facer, 2007, 2009; Van Dijk, 2005). However, most research focuses on the adult population and almost no attention has been paid to teenagers in general, and to young adults (between 16 and 24 years old) in particular, who lack access to the Internet. This is largely due to the widespread assumption that all young people belong to the digital native generation (Prensky, 2001), which is uniformly wired. This prevalent discourse implicitly spreads the idea that the digital natives are a homogonous age group, who are evolving in a digitally undifferentiated environment. Based on this assumption, a lot of research often fails to acknowledge that the benefits on being online are not distributed equally among all sections of youth. Such an interpretation is notably bolstered by a quick analysis of the results from quantitative surveys on young people's ICT uses. For example, in Belgium, the Statbel survey (2008) reports that only 9 percent of all 16-24-years-olds are rare Internet users or non-users (have not used at least for the last three months). 91% report to using the Internet at least *occasionally* (once a week), among whom 75% declare themselves to be assiduous users (almost everyday). From this perspective, it would probably seem to be more relevant to move attention away from digital inequalities among the digital natives rather than exploring the nuances of inequality in access and use of ICT within this one generation. Yet, they are even more exposed to a risk of digital marginalization since they only represent a minority within their generation. This angle of vision thus tends to ignore differences between youths with favourable and unfavourable conditions of access and use. Furthermore, it homogenises the online experience of the more as well as the less advantaged segments of the young adult population. In reality, however, the disparities in terms of socio-demographic characteristics among the 16-24-year-olds influence the heterogeneity of this group in terms of quality of access to the Internet, autonomy of its use and online expertise to a large extent (Livingstone, Bober, & Helsper, 2005; Livingstone & Helsper, 2007; Livingstone, 2008).

To contribute to remedying this lack in research, this chapter aims to examine the (subtle) forms of digital inequalities among young people between 16 and 25 years old in Belgium and examines what the consequences are in terms of social inequalities. It focuses on young adults who are rare or non-users of the Internet. Focusing on digital disparities among the 16-24-year-olds is all the more relevant since these (almost) *off-line* young adults are experiencing a series of transitions in their personal life and thus become gradually more concerned by the uses of the Internet in all areas of their life.

This article is based on an action-research carried out in 2009 by the Fondation Travail-Université for the Ministry of Social Integration, within the national action plan to overcome the digital divide (Brotcorne, Valenduc, & Mertens, 2009). The analysis draws on secondary analysis of data from quantitative surveys, and collective and individual interviews with social workers who are in charge of economically and culturally disadvantaged young adults in the three regions of the country – Brussels, Flanders and Wallonia.

The paper is divided into three sections: the first one will present a review of recent theoretical and empirical research regarding the digital divide and young people's digital uses of technologies, and considers a number of reasons why the problem of digital inequalities among the young population remains largely ignored by most research in this area. The second section describes the research method and the empirical data collected during the action-research. It exposes in particular how the field actors were involved in the empirical phase of this study. In the third section, findings from quantitative surveys and an analysis of focus group discussions are shown in detail, challenging the dominant discourse regarding the "digital native" generation and ICT. The article concludes by offering a set of research suggestions and policy recommendations to reduce the specific forms of digital inequalities that young adults have to deal with.

The Ambiguities of the Digital Natives Discourse

For the past fifteen years, increasing academic attention has been given to the issue of young people and digital technologies. Indeed, since ICT is an integral part of young people's daily life, a considerable body of research has focused on the relationship between youth and all these new technologies – computer, Internet, online games, mobile phone, MP3 – in order to analyse the nature of their digital practices.

A main transversal idea crosses the whole literature: youth are at the heart of the digital revolution. They are the first generation to be born with digital technologies and to have been immersed in a "culture of the screen" since childhood. They are all seen as intensive users. As a lot of academic research has shown, they massively adopted digital technologies across all domains of their daily life. New media consumption is thus one of the main activities of young people (Mediappro, 2007; Lenhart, Kahne, Middaugh, Macgill, Evans, & Vitak, 2008; Livingstone *et al.*, 2005).

In spite of many scientific investigations conducted on young people and ICT use, little research has yet focused on "offline young people" and the reasons why some young people make little or no use of the Internet. This is partly because young people are widely seen as *online experts* or *innate super users*. They were thus called *digital natives* (Prensky, 2001), *Internet generation* (Tapscott, 1998), or *generation C* (referring to characteristics such as being connected, creative and clicking) (Veen & Vrakking, 2006). In fact many terms have recently sprung up to name this young generation, born at the end of the 1980s and the beginning 1990s, who have intuitively learnt how to surf on the online information ocean. Evolving with ease in the digital world, they are seen as being able to simultaneously develop and handle various digital technologies. Someone has consequently qualified them as the *multitask generation* (Wallis, 2006). Since ICT characteristics seem to be perfectly adapted to young people's lifestyle (mobiles, usable in an autonomous way and able to quickly answer to their informational as well as their communication needs), digital technologies are seen as a ready means to integrate young people into the information society.

This enchanting discourse on a young generation uniformly connected to their mobiles and laptops quickly became popular. This has largely contributed to encouraging researchers to pay attention to topics regarding the dangers and the risks of ICT over-consumption by the digital natives, rather than on their opportunities in terms of social integration (for Belgium, see Bauwens, Pauwels, Lobet-Maris, Poullet, & Walrave, 2008). Moreover, most researchers have paid attention to the ICT use of children and teenagers from ages 11 to 16 or 18 years old, rather than that of young adults between 16 and 25 years old. In fact, few studies are addressing the issue of the digital behaviour of this latter age group marking the transition from adolescence to adulthood. However, it is precisely a group that experiences a series of transitions in personal life – from school to labour market, from family life to autonomous life – and they become increasingly concerned by the use of Internet in all areas of social life. Research has mostly approached ICT use from the point of view of online entertainment and communication rather than studies or work.

Due to this particular orientation of the debate, research attention often ignored digital inequalities in young people's access to and use of the Internet (Facer & Furlong, 2001; Livingstone *et al.*, 2005; Selwyn, 2009; Selwyn & Facer, 2009). Yet, as Neil Selwyn notes in a report drawn up for the Council of Europe:

> The issue of digital inequalities among young people should not be seen as being unproblematic [...] One cannot affirm straightaway that ICT uses give the same advantages to all the social groups of young people. If digital technologies can make certain young people more autonomous, they can also involve more marginalization of groups of young people more exposed to digital inequalities (Selwyn, 2007b, pp. 2 and 10).

It is then just as crucial to examine the problematic of digital exclusion among younger generation as among the adult population (Livingstone & Helsper, 2007).

Nuanced View of Young People's Internet Use

In this way, some recent and more empirically grounded studies have drawn a more moderate and complex picture of access to and uses of new technologies by the presupposed *digital natives* (Selwyn, 2007a, 2007b; Selwyn & Facer, 2007; Livingstone *et al.*, 2005; Livingstone & Helsper 2007; Hargittai & Hinnant, 2008; Selwyn, 2009).

These empirical studies take a critical perspective of the digital native discourse, and mostly focus on the reasons why some young users make little use of the Internet (although see Livingstone *et al.*, 2007). They do not, however, address the particular question of no ICT use at all by certain groups of young people or why these young people are totally offline (see in particular Hargittai & Hinnant, 2008). As Livingstone and Helsper suggested in a somewhat provocative way: "does one have to conclude that young people have no difficulties accessing and using the Internet or that inequalities do not divide them?" (Livingstone & Helsper, 2007, p. 672). The authors answered their own question with a firm negative. Together with other researchers, she assumes that it is increasingly difficult to identify one clear digital divide among this generation since very few young people can still claim today to be completely and continuously disconnected from digital technologies. To see the digital divide – access/no access and use/no use – as a simple dichotomy is too crude to understand the process (non-)engagement by young people and individuals with digital technologies as a whole. Instead, it is necessary to reconsider digital inequalities in a more subtle and nuanced way. A previous research for the ministry of Social Integration in Belgium (Brotcorne & Valenduc, 2008), underlined that, once the barrier of access is crossed, new disparities appear in ICT use among those who are already users. Formal access to ICT does not automatically lead to an autonomous and effective use of these technologies and their online contents. In this sense, though it is necessary to reduce inequalities in terms of access to ICT, mere access is insufficient to ensure equality of opportunities in usage. Indeed, concern for access has not yet resolved the question of what benefits Internet use brings exactly, and it is too often simply assumed that being online is necessarily a good thing. Thus, it is vital to examine disparities regarding the quality of access as well as the forms of engagement with digital technologies, not forgetting the nature/scope of their uses. This applies to young people no less, perhaps even more, than to adults. However, as Selwyn and Facer (2009) point out, these inequalities among the first group are less apparent than among the second one.

From this perspective, it is more relevant to consider digital exclusion among the young generation in a way that takes into account various inequalities at different levels. Livingstone and Helsper (2007) propose to consider the digital divide as a continuum, or a gradation, which goes from exclusion to digital inclusion among children and young adults.

Data and Method

In order to improve the policies of the national action plan against the digital divide in Belgium, the study carried out for the Federal ministry of social integration took the form of an action-research. To carry out this research, three complementary approaches were implemented. In addition to an overview of the scientific literature on digital exclusion among young people and an analysis of existing quantitative data (from public surveys and others), the methodology gave pride of place to an empirically grounded design. It requested the participation of field actors involved in associative or institutional initiatives for disadvantaged or vulnerable youth groups, in the three regions of Belgium – Flanders, Wallonia, Brussels. This interactive section was based on two workshops – one in Dutch, the other in French – with social and educational workers who were in charge of vulnerable young people in this age group. Initially we addressed youth organisations or youth services that had ICT training programmes for this age group. Nevertheless, in order to reach out to the target group, we very quickly broadened the interaction to a representative sample of the organisations and institutions in charge of wider initiatives for reaching those young people, regardless of their existing concern for digital exclusion within this age group. These focus groups were supplemented by a set of individual interviews with others field actors.

The methodological choice to organize workshops with the field actors working with young people, rather than the youths themselves, is due to time and financial constraints as defined by the Ministry of Social Integration. Although it would be very interesting to collect ICT youth's experiences directly, the voice of field actors became very relevant from the moment that the study's goal was to formulate political recommendations.

In confronting analyses from the existing research literature with the practices of field actors, the focus groups aimed to better understand which forms of e-exclusion young adults in transition have to face, in order eventually to formulate policy recommendations in line with the reality of the field. In sharing common actors' experiences, this qualitative method has the advantage of grasping both the agreement and the lines of dissension between the participants. Prepared by the research team, the interview guideline was used to direct the whole discussion.

Research Outcomes: A Nuanced Picture of the Digital Natives

Enlarging the Notion of "Offline Young Adults"

Very few young adults aged between 16 and 25 have never used the Internet: only 5% in 2008 according to Statbel survey (statistical national office of Belgium). These young people, called *offline youth*, are decreasing in number according to other Belgian statistical surveys, such as the annual survey of the Walloon Agency of telecommunications (AWT) on ICT use by the Walloon citizens or InSites marketing consultancy. According to Statbel (2008), 75% of 16-25-year-old people are daily Internet users, while 9% are non-users and rare users (not during the last three months); the remaining 16% are irregular users (less than once a week).

Compared to other European countries, the proportion of young non-users is the same as in France and Great Britain. This proportion is weaker in the Netherlands, Luxembourg and Germany, where the percentage of "completely offline" young people is insignificant.

Figure 1: Frequencies (%) of Internet use among young people (16-25-year-olds) in Belgium

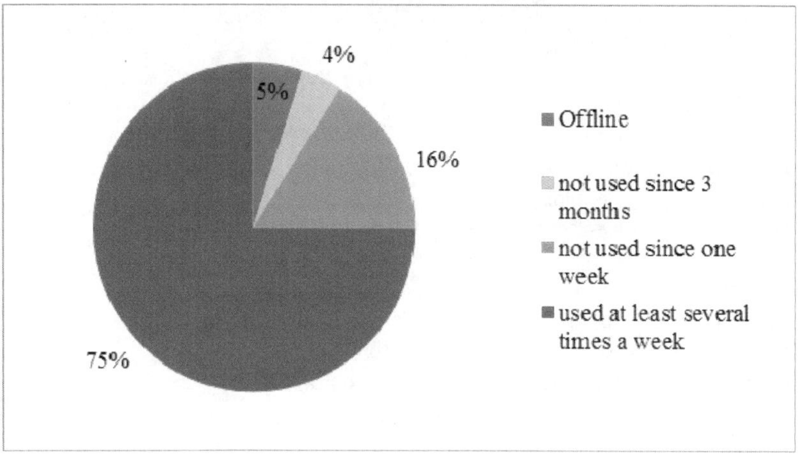

Source: Eurostat, Statbel 2008

Results from the focus groups with educators and fieldworkers largely confirm this trend. All participants agree in recognising that young people aged 16-25, who are completely disconnected from the digital world of today, are a marginal phenomenon in Belgium, even among the economically and socio-culturally vulnerable groups. As field actors explain, the vast majority of them have an opportunity of being connected anywhere, at least occasionally, to the Web via cyber cafés, youth associations, and any places of public access, or simply at a friend's home. Almost all young adults have access today to a basic Internet connection. All of the data suggest therefore that youth who never use the Internet are a marginal phenomenon.

Does this mean that young adults are all sheltered from digital inequalities? Certainly not. Conversely, the results suggest the necessity to consider digital inequalities in a more subtle way than as a simple binary divide and to extend the notion to include people's use of the Internet and the way digital media are incorporated into people's daily lives. Deeper discussions with the participants of the focus groups underline the fact that there is a wide range of situations of (quasi) disconnection, which may lead to various forms of social marginalization or exclusion, even once a youth has crossed the initial connectivity barrier and uses the Internet sporadically. Indeed, digital inequalities must not be understood as the simple result of the ICT use of some people and non-use of other ones. Rather, such inequalities result from the *outcomes* and the *consequences* of the uses of some of them in comparison with the low or the non-uses of the other ones, in terms of social discrimination. In this way, young adults who have a limited use of the Internet, in terms of frequency or types of usage, are also often excluded from many important online resources and services, which require being online on a daily basis. Online activities, such as text messaging, updating a blog, or updating a social network's page involve being regularly online at the risk of being rapidly sidelined. In other words, intermittent users are less likely to engage with social, capital-enhancing online activities than the daily users. The Internet fully plays its role of social and communication medium whether people use it on a regular basis or not. This is especially true for the young generation, for whom the Internet has an important identity-forming function in day-to-day life. In this perspective, regular users are likely to have types of online activities that bring social benefits whereas rare or irregular users tend to have fewer pay-offs. This distinction thus makes manifest new forms of digital inequalities.

Therefore, the reality of the Internet (non-) use by the young generation and of its consequences in terms of social identification with its own age group, calls into question the notion of *offline* young people *stricto sensu* (people who never use the Internet). It strongly suggests widening this notion in order to include situations of partial, intermittent and sporadic ICT disconnection.

Various Instances of Quasi-disconnection from the Internet

Young people, who are quasi-disconnected from the Internet are not exclusively a group with common socio-demographic characteristics. Whilst socio-demographic factors, such as socio-economic status among other things, are clearly important explanations of digital inequalities among young people and adults (Lenhart & Horrigan, 2003; Lenhart *et al.*, 2008), other contextual, cultural and psychological factors supplement and interact with these traditional ones to explain differential adoption and use of the Internet. Situations of quasi-disconnection cannot exclusively be attributed to the socio-economic background of young adults; they can also result from a lack of motivation or perceived usefulness, lack of skills or result from the perceived social influence of influential individuals in the environment, such as parents, family or friends with low e-culture (Broos & Roe, 2006; Hargittai, 2010).

Besides, the AWT survey (2009) gives a more nuanced picture of quasi-disconnected young people. According to this survey, intermittent and non-users are not characterized by unfavourable social conditions, neither in terms of standard of living, nor in terms of socio-professional category, nor in terms of their type of family. They have a low or medium level of education, yet they all have a high-tech phone and often a game console. Indeed, 16% of young adults who are following nonuniversity studies use the Internet less than once a week. Thus, these young adults are not strangers to the ICT world and they are familiar with the Internet, but they have other priorities and they probably do not need the Internet in their training courses.

Results from Broos' research (2003) also give a more nuanced picture of low and non-users of the Internet and demonstrate that psychological variables are essential in understanding the limited use of ICT among an adolescent population. Her explanatory model, which incorporated socio-economic characteristics as well as psychological concepts, reveals that self-confidence, lack of motivation in using ICT, outcome expectations and subjective norms are all key factors in the process of ICT (non-)adoption by adolescents. The extent to which they are confident about their ability to use digital applications, and the extent to which they like and make sense of using it are strong predictors of ICT adoption and use.

Discussions with educators and fieldworkers confirm that digital inequalities do not always and exactly map existing socio-economical divisions. In line with the previous research's results, participants of the focus group emphasised the fact that many young adults seem not to be interested and motivated in learning to use Internet applications that are not directly useful for them. According to the field actors, this lack of directly perceived usefulness comes from the difficulty many youths have in anticipating their future (professional) ICT needs and its future social status in general. Similarly, Broos' results showed that adolescents who believe they will attain a higher professional status are more likely to broaden their Internet usage than ones who do not believe it. It therefore demonstrates that anticipated social status, rather than current social status, is quite critical in explaining adoption or rejection and use or non-use of digital applications.

Furthermore, field actors also report the influence of bad schooling experiences on the limited use of the Internet. The impact of bad schooling experiences on ICT adoption partly operates through a perceived lack of self-confidence in learning situations. Young people who experienced these kinds of bad experiences have developed a bad individual judgment regarding their own skills and capabilities of learning to use the Internet. In this way, perceived self-confidence can have an important negative influence on the motivation to use digital applications more extensively. The influence of peer groups or of the family who are lacking in e-culture, reinforces this non-motivation, even fear, with respect to Internet use. The relationship of friends and family with the digital world operates as a subjective norm with which the youth identifies itself.

In sum, low or non-use of the Internet among the target age group is due to a wide variety of causes. Beyond the basic demographic and socio-economic factors, other dimensions, such as autonomy of use, skills, perceived usefulness and self-confidence, social and cultural support must also be taken into account to explain the differences in Internet adoption and use.

Different Worlds of Internet Usage

According to the quantitative and qualitative findings above, almost any young adult can be considered as totally disconnected from ICT. Except for some very marginalised young people, they are all, in one way or another, familiarised with the Internet. In this sense, digital inequalities within young people's groups take a particular shape: they are linked to the nature of ICT engagement.

Figure 2: Proportion of users of various online services according to age group

	Among the population (age group)			**Among Internet users (age group)**		
	16-24	25-34	35-44	16-24	25-34	35-44
Communication via Internet (all types)	88	76	71	97	93	90
Sending and receiving e-mails	85	74	70	93	91	88
Telephone via Internet, videoconferencing	14	16	11	16	19	14
Other types of communication (chat, social networks, etc.)	62	29	19	67	36	23
Download and use of music or video	53	28	23	58	34	29
Download of software	29	18	15	32	22	18
Use of web radio or web TV	30	19	13	33	23	16
Searching information on health or nutrition	26	31	32	28	38	40
Searching information on goods and services	68	70	68	75	85	85
Reading online newspapers or magazines	21	23	19	22	28	24
Interaction with public authorities (e-government)	19	32	28	21	40	35
Seeking for a job	13	16	9	14	20	11
Use of travel or accommodation online services	39	43	42	42	52	52
Use of online banking	31	48	45	34	59	57

% of Internet users within each age group – source: Statbel/Eurostat, 2008

The digital world of young people has common usage characteristics. As is shown in figure 2, the e-world of young people (16-24) differs radically form the practices of the succeeding age groups (25-34 and 35-44). While the youngest are more frequent users of the upper items of the graph (related to communication and entertainment), they generally use fewer information and transaction services than the older age groups. That gives this age group a contrasting profile in its Internet usage. Indeed, communication and entertainment activities (web 2.0. chat, games, music and film downloading, web radio/TV) are much more widespread among 16-24-year-olds than for both older groups. These last generations show a significant preference for other kinds of Internet use, such as information retrieval, administrative and commercial activities.

These findings are consistent with conclusions drawn from other, foreign research results. Beyond the disparities of young people's type of engagement with the Internet, quantitative and qualitative surveys of young people's uses of digital technologies from different countries nevertheless highlight common usage trends: as a whole, young people are likely to use the Internet for communication and entertainment activities above all. Results reveal a clear predominance of instantaneous text messaging, social networking, music and film downloading, game playing and retrieval of online content from YouTube, Facebook, Daily motion and other popular online services of the Web 2.0 (Mediappro, 2006; Lenhart *et al.*, 2008; Livingstone, 2009). While some commentators like to imagine collaborative and creative uses among young adults, in reality most of their engagement with ICT is limited to basic games and media consumption. One can thus argue that these specific usage patterns define the contours of a "young profile" in Internet use.

These common trends in young people's uses of the Internet are notable because the Web plays a major social and identity-forming role for this generation. Research results show that the Internet has a central place in the construction of common references for this generation; it is a digital medium that largely contributes to federalising young people's cultural practices (Pasquier, 2005). These functions of the Web are what distinguishes the digital practices of youth from those of adults today.

Another recent study by JOP (Boonaert & Vettenburg, 2009) also tends to frame our target generation's digital profile better. It shows that the existence of specific ICT uses does not mean that they are all necessarily homogeneous. In fact, it highlights four dominant clusters of uses among the young generation. According to this research, Internet can have an entertainment function; an information function; a communication function and a commercial function. Young people variously prefer each function of the web according to their age, their sex, and their parents' level of education.

Figure 3 schematises, in the form of a word cloud, the difference between the uses preferred by youths and the uses expected by the socio-economic world.

Figure 3: Word cloud of preferred uses by youths and uses expected by the socio-economic world

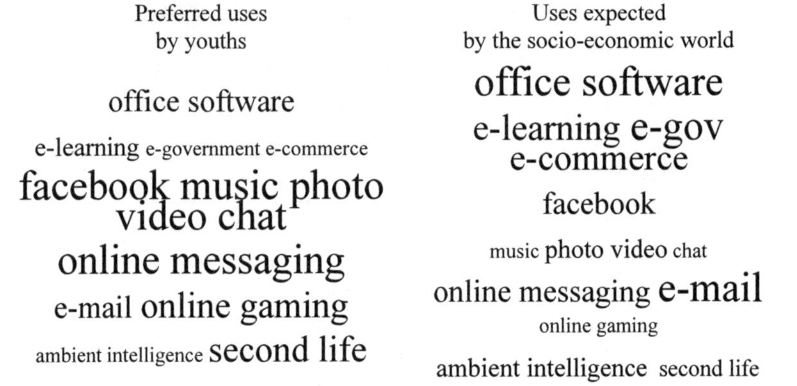

Focus group discussions with field actors have confirmed quantitative findings. Since almost nobody within the 16-24 age group can be considered as totally *offline*, digital inequalities have to be understood in another way. According to them, Internet inequalities within this age group are less related to access problems than to the nature of their Internet engagement. Basically, in this case, Internet inequalities have to be understood in terms of a disconnection between the digital experiences of youths – primarily focused on entertainment and communication activities – on the one hand, and the wider digital skills required for integration in the labour market and within society, on the other hand.

However, the e-knowledge of young people has to be seen as a continuum rather than in clear divisions. Discontinuities between both worlds are significant at both ends of the spectrum. At one end of the spectrum, one finds those who have material difficulties in accessing the Internet and have poor e-usages. They suffer a double e-exclusion: they experience difficulties in participating in the e-culture of youth and they lack enough competence to access the e-world of adult society. At the other end of the spectrum, we find a group of youth that can easily bridge the gap between their e-world and the *adult* e-world. Between these two extreme groups, one can find a much larger group of young people who are either not aware of the risk of e-exclusion or acknowledge their lack of e-competence in accessing the e-world as expected by the economy and society.

Internet-familiar or Internet-competent?

Although the majority of young adult are familiar with the computer and the Internet, their *e*-competencies and *e*-literacy remain less extensive than generally assumed. According to surveys, young adults often display a limited ability to successfully use all of the functions of the Internet and, even more, the computer. According to Statbel/Eurostat survey, whilst almost no young adults (16-24-year-olds) declare themselves unable to perform any task on the Internet, results indicate that 36% of them are only able to perform rudimentary tasks, such as using a browser, sending an email with attachment and so on. For more elaborate tasks, such as web page creation, less than 20% of them declared themselves able to perform it. Their e-abilities are even more limited with regards to the computer: 33% of young adults consider their competences as insufficient in facing the requirements of the labour market.

Many studies reveal empirical evidence that young people's current uses of digital technology remain more limited in scope than the digital native discourse leads us to believe. The prevalent Internet practices of young adults are relatively rudimentary. As pointed out by Livingstone in a recent study (2009), young people's engagement with the Internet is often far more passive, solitary and sporadic than the digital native discourse suggests. Young people's online searching strategies are also found to be relatively unsystematic and ineffective. They often display a limited ability in successfully using the Internet and other research tools (Williams & Rowlands, 2007). Although the *children of the Web* use online research more quickly than the older generation, they devote little attention to evaluating the quality of information, its relevance and its exactitude. Moreover, results also show that their ineffective online searching strategy is due to their lack of understanding of their own needs for information (JISC, 2008). A Belgian study (Pochet & Thirion, 2008), which evaluated the informational culture of students entering higher education, drew similar types of conclusion. Results notably underscored the fact that students seriously lack the critical literacy needed when reading web content and also lack competence in using search engines in an effective way.

However, other research results reveal that disparities related to the quality of use remain important amongst different youth groups. Young people's ability to make an autonomous and effective *use* of digital information is not automatic. For example, in a report on children and young people's Internet literacy, Livingstone *et al.* (2005) identified a range of socio-demographic barriers to Internet literacy and showed that the more Internet-literate young users are, the broader their uses of the Internet are and the more they are able to take up online opportunities. It would therefore appear that certain social groups of young people are much less likely than others to fully benefit from the access and use of these digital technologies (Dutton, Kahin, O'Callaghan, & Wykoff, 2005; Chinn & Fairlie, 2004; Holloway, 2005). In this sense digital inequalities regarding quality of use depend on a whole series of variables, closely related to certain demographic characteristics (sex, age, ethnic origin, geographical location, and handicap) and socio-economic characteristics (level of incomes, level of diploma, situation of school).

In sum, many foreign and Belgian research results point to common gaps and weaknesses in young adults' ICT practices: even if they have the functional expertise required to locate and acquire information – i.e. an operational know-how in finding and extracting online contents – they may lack the critical literacy that is required to interpret, understand, criticise and manage that information. In this respect, such empirical results tend to invalidate the myth of young adults as the online experts generation. In this perspective, the concept of digital divide takes on a particular meaning: it is related to the discrepancy between two digital worlds, which do not require the same kinds of ICT skills.

Focus group discussions with field actors largely confirm these empirical conclusions. Indeed, all participants recognise that the present young generation is using the Internet in a shared and specific way, beyond the different socio-economic and cultural environments they come from. They also agree that youth often display rudimentary and limited abilities in using the computer as well as the web. However, digital expertise is likely to be more extensive when young adults have a better socio-economic status and a higher level of education. For instance, as field actors explain, despite spending many hours on the Internet playing games, many vulnerable young people are lost when they have to search specific job-related or administrative information online. They also lack the ability to successfully fill in electronic job application forms and so on. Thus, contrary to common beliefs, a lot of public online information and many services are not used by young adults. This is particularly true for those who come from disadvantaged groups. For them, the discrepancy between both digital worlds is wider, as they lack parental or social support to help them to develop adequate e-skills regarding the requirements of society.

As focus group participants also stressed, these specific forms of digital inequalities remain underestimated by public authorities, labour market actors and education structures at different levels. This lack of knowledge regarding the mundane reality of young people's ICT use is largely due to the widespread idea that all young people belong to the digital native generation. Thereby few initiatives are taken by formal institutions (i.e. schools, labour market) and by educators to realistically support vulnerable young people in building bridges between the two digital worlds, in order to develop a meaningful engagement with ICT that might increase the empowerment of youth in contemporary society.

How to Bridge the Gap?

Over the last years however, research has gained a better insight into the way young people experience ICT. Conversely, schools and youth organisations still have little knowledge of the youth e-world. They are but poorly aware of the effects the Internet has on their socialising process and on the construction of their identity. Moreover, the idea that their e-competences might not be sufficient in order to compete on the labour market, rarely lead to measures that might facilitate the migration from their e-world to the one required by the socio-economic world.

Only the group of better-educated young people succeeded by themselves in operating the conversion from one e-world to the other. The majority of young people experience difficulties in their transition to the socio-economic world online and will need some kind of assistance. To bridge the gap between these e-worlds, the awareness of the difficulties that young people are experiencing in responding to socio-economic expectations must be significantly raised and efforts to take concrete action displayed.

The participants in the workshops favoured policies that would address the risks of e-exclusion that young people encounter when they migrate to the e-world of socio-economic actors. However, these measures will only be successful on condition that people who are in charge of youngsters within schools and the youth sector are trained to take up this role, which is not the case today.

It also becomes apparent that young people are deserting existing open-access infrastructures that provide the general public with an access to the Internet, mainly because they are ill-adapted to their demands. Open-access infrastructures should be more embedded in places where young people already meet, and where they are sufficiently confident to overcome their shame and ask for individual help when necessary. Finally, measures and methodologies for assisting young people in their migration to the e-world of the information society will only be successful on condition that they give more value to the existing e-competences of young people.

Conclusion

This paper focused on young adults between 16 and 25 years old – i.e. youth in transition between adolescence and adulthood – who are low or non-users of the Internet in Belgium. It examined the forms of digital inequality they have to cope with and its consequences in terms of social inequalities.

The review of academic literature, the findings from quantitative analyses and from the focus groups discussions converged in acknowledge that the number of young adults who are completely offline is very restricted, so that they are quite a marginal phenomenon. These young adults represent a very heterogeneous social group rather than a group with common socio-demographic and socio-economic characteristics. Thereby, these findings strongly suggest widening the notion of offline young people, as initially defined, in order to include situations of partial, intermittent and sporadic ICT disconnection.

Surprisingly, quasi-offline young people do not come strictly from socio-economically and culturally disadvantaged households. The reality of digital inequalities among the young generation appeared to be more nuanced and complex than was generally considered. Whilst socio-economic barriers remain important factors with respect to digital inequalities, especially in terms of quality of access, other reasons (psychological and cultural barriers, experiences with the Internet, perceived direct irrelevance of Internet use, lack of self-confidence and social support) interact with the traditional socio-demographic barriers to explain differentiated levels of access and use of online contents among this group, even once they had crossed the basic gap of being connected at all.

In this perspective, in line with a qualitative model designed in Broos' research (2006), it is possible to discern within our empirical data the contours of different, but not exhaustive, types of rare or intermittent Internet users: the first – minor or even marginal – group are the totally excluded who have no access to ICT at all for diverse reasons. The second group consists of those young adults who are subjected to strong control or norms by parents or peers as regards the Internet. The third group is made up by the indifferent, those who do not often use the Internet because of a general lack of motivation, whether due to a lack of perceived direct usefulness of digital applications or a lack self-confidence in their ability to use it.

More generally, although almost all young adults, even those who come from disadvantaged groups, are familiar with the Internet, the digital inequalities that remain take another, specific form. They are related to the discrepancy between the specific ICT uses and competences of youth – mainly orientated to passive media and games consumption – on the one hand, and the digital skills required by public authorities, education structures and the labour market on the other hand. Therefore the risks of e-exclusion concern young people who have a limited use of the Internet and are lacking social support to migrate successfully from their own e-world to the e-society as expected by the social norm of the information society. These youngsters, who gain few social benefits from their occasional online activities, need appropriate social intervention that supports their "migration process" because they are unable, until now, to do so autonomously.

This conclusion points out the importance of taking a more nuanced view when studying the relationship between Internet use and social inequality. Far from being simply dependent on mere access, systematic differences are present in how people incorporate digital media into their lives, even once they have crossed the first "connectivity" barriers. These differences have the potential to perpetuate inequalities among the "connected".

References

Bauwens, J., Pauwels, C., Lobet-Maris, C., Poullet, Y., & Walrave, M. (2008). *Cyberteens, Cyberrisks, Cybertools – Les teenagers et les TIC, risques et opportunités*. Rapport TIRO pour le programme Société et Avenir de la Politique scientifique fédérale.

Boonaert, T. & Vettenburg, N. (2009). *Jongeren en ICT: een divers publiek*. In N. Vettenburg, J. Deklerck, & J. Siongers (eds.), *Jongeren binnenstebuiten*. Leuven: Acco.

Boutet, A. & Trémembert, J. (2009). Mieux comprendre les situations de non-usage des TIC – Le cas de l'informatique et d'Internet. *Les Cahiers du Numérique*, *5*(1), Hermès Lavoisier.

Broos, A. & Roe, K. (2003). The Digital Divide in the Computer Generation: ICT Among Adolescents. Paper presented at "The Digital Dynamics Control, Participation and Exclusion" conference, Loughborough, November 2003.

Broos, A. & Roe, K. (2006). The Digital Divide in the Playstation Generation: Self-efficacy, Locus of Control and ICT Adoption Among Adolescents. *Poetics*, *34*(4-5), 306-317.

Brotcorne, P. & Valenduc, G. (2008). *Construction des compétences numériques et réduction des inégalités*. Brussels: Rapport SPP Intégration sociale.

Brotcorne, P. & Valenduc, G. (2009). Les compétences numériques et les inégalités dans les usages d'Internet. *Les Cahiers du Numérique*, *5*(1), Hermès Lavoisier.

Brotcorne, P., Valenduc, G., & Mertens, L. (2009). *Les jeunes offline et la fracture numérique. Les risques d'inégalités dans la génération des natifs numériques*. Brussels: Rapport pour le SPP Intégration sociale.

Chinn, D. & Fairlie, R. (2004). *The Determinants of the Global Digital Divide: A Cross-country of Computer and Internet Penetration*. eScholarship Repository. University of California.

DiMaggio, P., Hargittai, E., Celeste, C., & Shafer, S. (2004). Digital Inequality: From Unequal Access to Differentiated Use. In K. Neckerman (ed.), *Social Inequality* (pp. 355-400). New York: Russel Sage Foundation.

Dutton, W., Kahin, B., O'Callaghan, R., & Wykoff, A. (eds.) (2005). *Transforming Enterprise: Economic and Social Implications of Information Technology*. Cambridge, MA: MIT Press.

Facer, K. & Furlong, R. (2001). Beyond the Myth of the "Cyberkid": Young People at the Margins of the Information Revolution. *Journal of Youth Studies*, *4*(4), 451-469.

Hargittai, E. (2002). Second-order Digital Divide: Differences in People's Online Skills. *First Monday*, *7*(4), University of Illinois at Chicago, April 2002.

Hargittai, E. (2010). Digital Na(t)ives? Variation in Internet Skills and Uses Among Members of the Net Generation. *Sociological Inquiry*, *80*(1), 92-113.

Hargittai, E. & Hinnant, A. (2008). Digital Inequality: Differences in Young Adults' Use of the Internet. *Communication Research*, *35*(5), 602-621.

Holloway, D. (2005). The Digital Divide in Sydney: A Socio-spatial Analysis. *Information, Communication and Society*, *8*(2), 168-193.

JISC (2008). *Information Behaviour of the Researcher of the Future*. Commissioned by the British Library and JISC (Joint Information Systems Committee).

Lenhart, A. & Horrigan, J.B. (2003). Re-visualising the Digital Divide as a Digital Spectrum. In *IT & Society*, *1*(5), 23-39.

Lenhart, A., Kahne, J., Middaugh, E., Macgill, A., Evans, C., & Vitak, J. (2008). *Teens, Video Games, and Civics*. Pew Internet and American Life Project.

Livingstone, S. (ed.) (2008). *Theorising the Benefits of New Technology for Youth: Controversies of Learning and Development*. ERSC Seminar Series on the Educational and Social Impact of New Technologies on Young People in Britain. University of Oxford and London School of Economics.

Livingstone, S. (2009). *Children and the Internet: Great Expectations, Challenging Realities*. Cambridge: Polity.

Livingstone, S., Bober, M., & Helsper, E. (2005). *Internet Literacy Among Children and Young People: Findings from the UK Children Go Online Project*. London: LSE. Retrieved June 1, 2010, from http://www.children-go-online.net.

Livingstone, S. & Helsper, E. (2007). Gradations in Digital Inclusion: Children, Young People and the Digital Divide. *New Media and Society*, *9*(4), 671-696.

Mediappro (2006). *The Appropriation of New Media by Youth*. European Commission, DG Information Society and Media. Retrieved June 1, 2011, from http://www.mediappro.org.

Mertens, L., Brotcorne, P., & Valenduc, G. (2010). Offline Youth and the Digital Divide: Revisiting the Concept of the "Digital Natives". In *Social Media for Social Inclusion of Youth at Risk*, Proceedings of the Incluso 2010 conference (pp. 73-80). Leuven: K.U.Leuven.

Messin, A. & Jouet, J. (2005). Jeunes internautes avertis ou l'ordinaire des pratiques. In B. Conein, F. Massit-Foléa, & S. Proulx (eds.), *Internet: une utopie limitée. Nouvelles régulations, nouvelles solidarités* (pp. 121-138). Presse de l'Université de Laval.

Pochet, B. & Thirion, P. (2008). *Enquête sur les compétences documentaires et informationnelles des étudiants qui accèdent à l'enseignement supérieur en Communauté française de Belgique*. CIUF.

Prensky, M. (2001). Digital Natives, Digital Immigrants. *On the Horizon*, *9*(5), 1-5.

Selwyn, N. (2006). Digital Division or Digital Decision? A Study of Non-users and Low-users of Computers. *Poetics*, *34*, 273-292.

Selwyn, N. (2007a). *Digital Inclusion for Young People: Ensuring Opportunities for Vulnerable Groups*. Remarks to Westminster eForum Keynote seminar: *UK Kids Online: Young People, Media Literacy and the Digital Divide*, 24th April, London.

Selwyn, N. (2007b). *Les jeunes et leurs besoins d'information dans le cadre de la société de l'information*. Bruxelles, Direction de la jeunesse et du sport du Conseil de l'Europe/Agence européenne pour l'information et le conseil des jeunes.

Selwyn, N. (2009). The Digital Native: Myth and Reality. Paper presented to *CILIP London seminar series*.

Selwyn, N. & Facer, K. (2007). *Beyond the Digital Divide: Rethinking Digital Inclusion for the 21st Century*. London: FutureLab.

Selwyn, N. & Facer, K. (2009). Beyond the Digital Divide: Towards an Agenda for Change. In E. Ferro, Y. Dwivedi, R. Gil-Garcia, & A. William (eds.). *Overcoming Digital Divides*. Hershey PA: IGI Global.

Selwyn, N., Gorard, S., & Furlong, J. (2005). *Adult Learning in the Digital Age*. London: Routledge.

Stevens, F. (2004). Jeugd en media. Media als verlengstuk van het dagelijks leven. In D. Burssens *et al. Jeugdonderzoek belicht. Syntheserapport van wetenschappelijk onderzoek naar Vlaamse kinderen en jongeren*. K.U. Leuven, VUB & UGent.

Tapscott, D. (1998). *Growing Up Digital: The Rise of the Net Generation*. New York: McGrawhill.

Van Dijk, J. (2005). *The Deepening Divide: Inequality in the Information Society*. London: Sage.

Veen, W. & Vrakking, B. (2006). *Homo Zappiens. Growing Up in a Digital Age*. London: Network Continuum Education.

Wallis, C. (2006). The Multitasking Generation. In *Time*, 167, 48-55.

Williams, P. & Rowlands, I. (2007). *Information Behaviour of the Researcher of the Future. The Literature on Young People and Their Information Behaviour. A British Library/JISC Study*. Retrieved June 1, 2011, from http://www.ucl.ac.uk/infostudies/research/ciber/downloads/GG%20Work%20Package%20II.pdf.

Statistical Sources

AWT (2009). *Usages d'Internet par les citoyens wallons – Enquête 2008*, Agence wallonne des télécommunications, Namur, juin 2009.

Eurostat (2009). *Base de données statistiques sur la société de l'information.*

Statbel (2008). *Statistiques sur l'usage des TIC par les individus*, SPF économie.

Statbel (2008). *Statistiques sur l'usage des TIC par les ménages*, SPF économie.

Part II

Content

Young People as Consumers and Producers

The Social Meaning of Young People's Online Creativity

Joke BAUWENS & Chris VLEUGELS

Introduction

If one were allowed to pick one promise from all the high hopes that have been circulating about young people and ICT, then the expectation of an enhanced, enlarged and empowered creativity for young people seems to be one of the most appealing. Not only scholars in cultural studies and critical media pedagogues, but also many policymakers, educational professionals, cultural institutions (e.g. public services broadcasters, museums) and industries seem to share the idea that a creative use of new media technologies is vital to the development of young people, both as cultural consumers and as cultural citizens (e.g. Goodman, MacCallum-Stewart, & Munsell, 2009, p. 44; Burgess, 2006; McDermott, 2008; Sefton-Green & Soep, 2007). In particular, the opportunities for productive creativity, resulting in novelty, innovation, and artistic expression, are highly valued. Equally, the theories and debates about creative industries, new forms of economy and entrepreneurship, as well as new forms of learning, seem to have converged in a strong belief in the creative potential of ICT and its young users for today's society (Tacchi, 2004; Hermes & Janssen, 2006; Burgess, 2006).

However, facts and figures from various surveys across Europe and the US show that the so-called "Generation C is not ubiquitous" (Pruulmann-Vengerfeldt, Kalmus, & Runnel, 2008; n. pag.). In these surveys the conceptualisation, operationalisation and measurement of digital creativity often vary, as some indicators about digital creativity are more about digital literacy or skills (what young people say they are capable of doing) than about creative practices (what young people report as actual practices). Still, international surveys over the last ten years suggest that productive creativity is less prevalent than consumptive creativity. Hence, the group of young people that uses the Internet for creating artefacts (e.g. developing a website, building a personal website, starting a blog, uploading user-generated content) is still smaller than the group that does not show these kind of practices (see Ofcom, 2006; Mediappro, 2006; Pruulmann-Vengerfeldt, Kalmus & Runnel, 2008; Bauwens, Pauwels, Lobet-Maris, Poullet, & Walrave, 2009;

Livingstone, Haddon, Görzig, & Ólafsson, 2010). This is also demonstrated in the EU Kids online survey, one of the most recent large-scale European studies on young people's use of online technologies. Receiving content produced by others (e.g. 83% watches video clips), playing games (e.g. 74% plays against the computer), and communicating (e.g. 61% uses instant messaging) are far more common than posting images (38%) or messages (31%), creating something in a virtual world (18%), or writing a blog (10%). These findings put the buzzwords or, as Schlesinger (2007, p. 377) states, the *hegemonic terms* used in today's discourses about young people, creativity and participation in another, eye-opening perspective.

At the same time, studies show that with the rise of the so-called Web 2.0, which is built on participation, collaboration and sharing, the social meaning of digital creativity has taken on new dimensions. Many young people are engaged in diverse online communication activities where they exchange, receive, re-interpret and rework various symbolic meanings. In today's age of converging media forms, ICT is pre-eminently mobilised by young people to construct a space for communication, community and sociality. This is shown in the continuation of offline sociality when going online, but also in young people's testing of their social identity, social belonging and different modes of social interaction amongst peers (see, among many others, Holloway & Valentine, 2003; Buckingham, 2008; Ellison, Steinfield, & Lampe, 2007; Mesch & Talmud, 2010).

The social networking sites (SNS) in particular have become important spaces where young people get connected, not only on the basis of their shared interests and roots, but also on the basis of creating content. Both through productive and consumptive creative practices young people develop a strong sense of social identity and social community (Burgess, 2006; Bloustien, 2007; Stern, 2008). Especially in reworking, re-mediating and recombining symbolic material provided by the cultural industries, or even more simply, in downloading and uploading this material on SNS, young people are communicating and expressing themselves for a public, both inside and outside their inner circle of peers, friends and relatives. The coming together of user-generated content and social software (see Burgess, 2007) seems to affect the Internet practices of many young people (Livingstone, 2008; Lenhart, Madden, Rankin Macgill, & Smith, 2007; Burgess, 2006). Today, community and communication are considered to be crucial constituents of creativity (e.g. Loveless, 2006).

All these developments and processes constitute the background of our study into the social value of young people's digital creativity. When we talk about digital technologies, we refer to the wide array of

ICT that young people use to exchange symbolic communication. Obviously, the Internet, home computer, digital camera, mobile phone and games play a pivotal role in these processes, mostly in a convergent way. Although our research goes back before the big breakthrough of SNS like Facebook, other Internet-enabled social communicative spaces, which are still prevalent today, are considered (e.g. instant messaging, e-mail, Myspace).

Along the lines of ethnographic methodology, in our research we started from the following open question: what social value does digital creativity have for young people? How are digital creativity and sociality among young people intertwined? How does digital creativity build and sustain social bonds among peers? What are for teenagers the social incentives and rewards for being creative on the Internet? Based on findings drawn from a qualitative research panel we explored these questions with 17 teenagers, aged 12 to 18. The research process involved, as Hammersley puts it: "both deduction and induction in the broad sense of those terms" (Hammersley, 1992, p. 48). So we combined two strategies in our analysis, "the widely read and the 'blank' one" (Alvesson & Sköldberg, 2000, p. 251), or preconceptions and *tabula rasa*, between which we alternated in order to understand what it means socially to be digitally creative from a young person's perspective. This double movement produced findings that resonate with the conclusions of other researchers, as referred to in the discussion of our findings, and also bring in new perspectives on the ordinariness and intimacy of digital creativity.

Creativity Is Ordinary

A key concept in the discussion around the potential of new technologies in young people's everyday life, digital creativity has been defined and operationalised in varied ways. One of the persistent areas of tension touches upon the question whether the creativity that young people display when making use of ICT is, on the one hand, extraordinary, exceptional and peripheral or, on the other hand, ordinary, mundane and ubiquitous (Stern, 2008).

The latter, more inclusive notion of creativity, as the *little* or *mini-c creativity* (see Craft, 2000 in Loveless, 2006, p. 8; Beghetto & Kaufman, 2007), aka *everyday creativity* (Pachucki, Lena, & Tepper, 2010, p. 123; Sanders, 2005), originates from the psychology and philosophy of pragmatism. As Pachucki *et al.* (2010) argue, from a psychological and pragmatist point of view, to live and to survive is by definition to be creative. This idea also strongly takes shape in the work of Sanders (2005) who distinguishes four levels of everyday creativity. *Doing* is the most basic level of creativity and is about getting something done.

Adapting is more advanced, as the motivation behind it is to make something one's own by changing it in some way. *Making* is about using one's hands and mind to make or build something that did not exist before. *Creating* is the most advanced level of creativity and refers to expressing oneself or to innovate. All four levels refer to the activities and skills people demonstrate on a daily basis when coping with the world and social life.

This everyday notion of creativity has established itself in media and cultural studies through the work of Willis (1990). Willis was one of the first to state that young people's engagement with popular culture, such as music and media, is an act of creativity. According to him, creativity should be understood as something that is engendered from the synthesis between consumption and production of popular culture, and is a more over-arching form of, what he called, *symbolic creativity*. Assuming that "young people are all the time expressing or attempting to express something about their actual or potential cultural significance" (Willis, 1990, p. 1), creativity through popular media culture was believed to be mundane and constitutive of the subjectivity or identity formation of young people (Sefton-Green & Buckingham, 1998, p. 62-63).

Willis, and many other contemporaries, above all Fiske (1990), enlarged the notion of media creativity to include the interpretative work people are engaged in when they consume popular culture. Rather than emphasizing the productive meaning of creativity, the one resulting in artefacts, many scholars working within the field of cultural studies or who have affinity with it, have argued that "mere media consumption" is also a creative and productive act as people are involved in meaning-producing activities. In thinking, reflecting, imagining, talking and discussing about popular media texts, even if not of their own making, people actively produce sense (Barker, 2008, p. 50-51).

Judging from the work and talk about what digital technologies can mean to young people's creative powers, this kind of creativity has been pushed into the background. As the accessibility, availability and simplicity of PCs, digital cameras, user generated content software and online mobile devices have improved for large, though not all groups in society (see Livingstone *et al.*, 2010), the creativity of youth has been much more defined in terms of production rather than in terms of consumption, as in music making, magazine production, video making, creative writing on blogs, video game development, etc. (Sefton-Green & Soep, 2007; Sefton-Green & Buckingham, 1998, p. 62; de Leeuw & Rydin, 2007).

Many studies show that digital creativity is indeed productivity, and that in experimenting with various ICT, from games to mobile phones, "opportunities for new kinds of artistic expression" (Sefton-Green &

Soep, 2007, p. 835) are created. However, a new hierarchical distinction emerges within the notion of digital creativity. On the one hand, there is a latent idea that the production of new, novel and original material (as in producing new texts, creating original music and video productions, designing games and virtual architectures) especially, produces aesthetical, artistic and original value (Niesyto, Buckingham, & Fisherkeller, 2003; Stern, 2008). On the other hand, many argue that the re-use of existing material (as in manipulating, remixing, re-mediating and re-combining) gives way to other aesthetic experiences. To these authors, *collage creativity* (Drotner, 2008), *bricolage* and *intertextuality*[1], *dialogic communication* (Buckingham, 2003, p. 134-135), or *produsage* (Bruns, 2005; 2008) are important sites of artistic, aesthetic and creative expression, and have become a constituent part of young people's everyday life (Burgess, 2006; Stern, 2008). In this context, we strongly sympathise with the concept of *vernacular creativity*, defined by Burgess as:

> [...] the process by which available cultural resources (including both "material" resources – content, and immaterial resources – genre conventions, shared knowledges) are recombined in novel ways, so that they are both recognizable because of their familiar elements, and create affective impact through the innovative process of this recombination (Burges, 2006, p. 6).

Informed by these approaches, the definition of the digital creativity we use in our study is in line with the most accessible one: a digital creativity that does not necessarily result in original artefacts. We believe that a more exclusive definition of creativity (as aesthetically distinguishing and innovating (see, for instance, Buckingham's discussion (2003, pp. 124-125)) devalues the practices young people actually (and not ideally) are engaged in. Although it is clear that the resources and structures through which teenagers are expressing themselves are predominantly provided by the cultural industries, and that many of them are operating within these margins, it is "the art of making do" (Fiske, 1990: 4) that seems to be the most fitting definition of youth digital creativity (see also Buckingham, 2003; Drotner, 2008).

Creativity Is Social

Building on sociological theory and empirical research into young people's ICT practices, it appears that the social meaning of creativity can be understood as the dynamic interaction between confirmation and distinction, two processes that give shape to the way social groups or, in

1 The notion of bricolage was first used by Claude Lévi-Strauss and introduced in cultural studies through the work of John Clarke and Dick Hebdige. The concept of "intertextuality" originates from Roland Barthes and found its way into the field of cultural studies mainly via John Fiske's writings (see Barker, 2008).

Bourdieu's (1979) terms, social fields are structured and organised, within, but also in relation to, other social fields. For the particular group and form of creativity we are studying, i.e. young people's digital creativity, these processes are closely intertwined with, first and foremost, the role of friendship and peer culture in young people's daily life (see Crockett, Losoff, & Petersen, 1984). Following De Block and Buckingham (2007), an important aspect of media productions by young people is to consider them in relation to their identity as friends. Media are part of the social life of young people, and through producing media, friends are not only connecting on a social level, they are also making sense of media in their lives.

Creativity is a social process as it often finds expression in groups and processes of social interaction where social norms, conventions and pressures are prevalent. As social beings we orient ourselves around shared identities and values, also when we are creative. This means, as Pachucki *et al.* (2010) argue, that creativity is an outstanding example of the tensions between social confirmation and social distinction. Creativity is always the combination of producing something "novel (i.e. original, innovative) and appropriate (i.e. useful)" (Pachucki *et al.*, 2010, p. 123). Hence, creativity is simultaneously about creating something distinctive and at the same making sure that the creation is somehow socially accepted. Burgess' definition, as quoted above, also makes reference to this area of tension. On the one hand she acknowledges the importance of recombining elements in a novel way, but at the same emphasises the role of the "recognizable" and the "familiar" (Burgess, 2006, p. 6). Because of the role of friendship in young people's everyday life it can be assumed that the norms and values peers share in their circle of friends are of importance to the creativity shown in the social group they belong to.

The play of social confirmation and social distinction is very evident in the way young people interact and present themselves through ICT. Youth studies and media-based educational projects have pointed out that digital media creativity among young people is a social process that inevitably entails social negotiations between peers, explorations of the relationship between the self and the others and staging of social identities through gender, ethnicity and class. In all these processes the boundaries of the acceptable and the normal are explored and negotiated (Buckingham, 2003, p. 128-130; Drotner, 2008; de Leeuw & Rydin, 2007). As Stern (2008, p. 108-109) concludes from her study, the drive to bond with peers, the desire for social and self-acceptance, validation and feedback, again from peers, about the appropriateness of who they are and what they do, is what impels young authors to share and publish their own symbolic content on the Internet.

This brings us to the second social component of digital creativity: the opportunity to go public and, what is more, to potentially reach a global audience. Studies on young people's online creativity show that the awareness of publicity plays a decisive role in the online practices they engage in and that the valorisation of publicity is often conceived of as an end in itself. As part of a larger spirit of the times, which cherishes the cultural values of self-promotion, young people are eager to express their identity online (Stern, 2008; Sima & Pugsley, 2010). Obviously, a substantial part of young people's explorations of social identity and social belonging takes place in public repositories, where many others, depending on the openness/closedness of the virtual spaces they are engaged in, can observe their playing with, rehearsal of, and training in making sense of themselves, of others and of the world they live in. On SNS, personal web pages, in gaming environments, etc. they present and represent themselves through published texts in various forms (written texts, pictures, videos, avatars drawings, music, speeches, etc.). These public performances can be viewed as "portals for the negotiation of social positioning and relics" of teens' identities (Dowdall, 2009, p. 77). Hence, teenagers not only try to manage the impressions they give of themselves, aware as they are of their audience, regardless of its size, but they also try to achieve a certain public status (Papacharissi, 2002; Donath & Boyd, 2004).

A common fallacy in popular discourses on Web 2.0 and its young users is that the high intensity the younger generations display in their peer-to-peer communication activities will, or, as some are claiming, does already coincide with a widespread habit to publish and distribute user-generated content to a global audience. However, as stated earlier, young people's inclination to publish their own symbolic content remains rather modest. High investments in what Castells (2009, p. 54) denotes as "interpersonal communication" do not necessarily result in busy "societal communication" activities. Or, to borrow Castells' language again, it is not because young people are interacting with each other with high frequency and at high speeds, that they feel the need or have the intention to send self-generated messages to many anonymous people, also known as the audience. This is the paradox of the Internet use of many young people. They communicate among themselves massively and heavily on the Internet, so a lot of their online communicative practices can indeed be denoted as "mass self-communication" (Castells, 2009, p. 55), but they themselves do not necessarily aim to reach a global audience. Rather, they see the virtual spaces in which they dwell as relatively private and intimate spheres because of the social manageability of these spaces (e.g. lists of friends) and because of the people they meet there (often peers) (Bauwens *et al.*, 2009).

Although many young people share their creative output and participate in conversations initiated by that content, just as many young content creators do not publish their creative performances on the Web for anyone to view and limit access to content (cf. Lenhart *et al.*, 2007). Stern's (2008) study reveals the feelings and ideas behind this widespread reserve with respect to mass self-publication. She concludes that young people's longing for being known, the correlate of the contemporary celebrity- and self-promotion culture, is often coupled with feelings of vulnerability when publishing self-made material. Knowing and hoping that others will encounter their online expression is the fundamental appeal of publishing online. Adolescents who never had many opportunities for public expression, may feel especially empowered by the prospect of being heard and seen by a global audience. At the same time, Stern points out in her study that nearly all the young people she talked to identified other young people as their principal audience (Stern, 2008, p. 103-108).

Research Method and Material

In order to reflect on the social value of young people's online creativity, we draw upon the qualitative data that was collected within the Belgian research project TIRO (Teens and ICT: Risks and Opportunities).[2] In this study we examined how Belgian teenagers (age 12 to 18) make use of ICT, how their usage relates to their social background, and how they experience ICT in different contexts and practices of their everyday life. The research consisted of a large-scale survey of 1,318 teenagers and 571 parents, and a panel research of 34 Belgian teenagers. In this paper we use the data from the Dutch-speaking panel that consisted of 17 teenagers.

The seventeen participants were selected through theoretical sampling. Based on a review of the (Western) literature, a set of social characteristics emerged as being qualifying for young people's use of the Internet. These characteristics were gender and age, and the levels of education and occupation of both parents (as determinants of the family's socio-economic status). In addition to the social characteristics, and to end up with a diverse sample, we also took into consideration the teenagers' (use of) technological equipment, their education and leisure activities, and the family structure. The Dutch-speaking panel consisted

2 The TIRO research project was funded by the Belgian Federal Science Policy Office. The following research groups were involved in the TIRO research project (February 2006-December 2008): IBBT-SMIT (Vrije Universiteit Brussel – Free University of Brussels), CITA (Facultés Universitaires Notre-Dame de la Paix de Namur – University of Namur), MIOS (Universiteit Antwerpen – University of Antwerp) and CRID (Facultés Universitaires Notre-Dame de la Paix de Namur – University of Namur).

of 8 boys (two aged 12-13; three aged 14-16; three aged 17-18) and 9 girls (three aged 12-13; three aged 14-16; three aged 17-18) as displayed in the table below.

In accordance with the methodology of ethnography (see Walsh, 2004 and others), we set up a panel study that consisted of a variety of techniques. During ten months, from November 2006 till August 2007, we had regular contact with the participants by means of face-to-face, in-depth interviews, instant messaging and e-mail, and participant observations in their home environment. In addition, we let them report about their use of ICT in a diary and looked at the content they published online (e.g. personal webpages). We also took into account technologically mediated creative practices that were not published online but were shared with us (such as photos). These forms of creativity are still interesting with respect to our research questions, especially in order to gain insight into what material got published and what did not.[3]

While the in-depth interviews and the diaries were plainly elicited, the conversations through instant messaging and e-mail were more spontaneous and informal. Creativity was one of the many aspects of the participants' digital practices we encountered and investigated. Obviously, as we aimed to study the natural engagements of young people with ICT, we did not stimulate nor provoke digital creativity among the participants. For instance, we never asked them to create and develop online profile pages, neither did we ask them to produce videos and publish them online.

Diversity of Creativity

Starting from an inclusive definition of digital creativity, in which young people engage with symbols provided by the culture industry and draw upon these symbols to create and express themselves, whether or not they are shared with others, all 17 teenagers from our panel could be described as creative. As shown in the table below, we could identify at least one creative practice for each teenager. However, as pointed out in other studies, the young people themselves did not necessarily consider their daily creative activities as creative, often because they had also internalised "the heroic notion of creativity that delimits activity to the fields of artistic and scientific activity" (Pachucki *et al.*, 2010, p. 138). In our research, creative practices such as contributing to a forum, writing status updates or adding photos or comments to a social networking site, were all seen as part of daily interpersonal communication

3 Our study was conducted before the breakthrough of Facebook (in 2008). During our research the market for social networking websites was still dominated by Microsoft's MSN-space.

and as such not recognised as involving creativity, even though these practices are definitely more than simply *doing* things, to borrow the language of Sanders (2005).

Table 1: Overview of the creative practices of the panel members

Age group	Gender	Name	Creative practices
12-13	Boys	Eric	Videos (not published) Role-playing games (Legend of Zelda, Warhammer fantasy)
		Frank	Drawings made on computer (not published)
		Vince	MSN-space Role-playing game (World of Warcraft)
	Girls	An	MSN-space, High (www.high.be)
		Diane	MSN-space Simulation game (SIMS)
		Shannon	MSN-space, Netlog (www.netlog.com)
14-16	Boys	Lewis	MSN-space, Netlog
		Nicolas	MSN-space, Myspace (www.Myspace.com)
	Girls	Amy	MSN-space
		Catherine	MSN-space PowerPoint as gift for an ill friend
		Lisa	MSN-space, Myspace Forum of a Belgian rock band
17-18	Boys	Tom	MSN-space Videos (published on YouTube) Role-playing game (World of Warcraft) Forum on gaming Compilation-CD as a gift for Valentine's Day
		Alex	MSN-space Videos (published on YouTube) Simulation game (MegaTycoon)
		Justin	MSN-space Role-playing games (World of Warcraft, developed a game as dissertation project)
	Girls	Eve	MSN-space Video (not published) Magazine as gift for a birthday
		Julia	MSN-space
		Clare	MSN-space, Partypeeps (www.partypeeps2000.com), Myspace, Slide (www.slide.com) Wikipedia

Note: Since all participants were under 18 years old at the time of the study, we decided to substitute pseudonyms for the real names of the teenagers. However, not a single participant asked us to disguise his or her identity in our publications.

Looking at the table, showing our overview of young people's digitally creative practices, it can be gleaned that the most basic level of creativity, i.e. *doing*, is missing. Obviously, this form of creativity is hard to grasp and asks for a closer scrutiny of daily activities. At the same time, we were surprised about the ubiquity of the more advanced levels of creativity, *adapting*, *making* and *creating* (see Sanders, 2005).

With the exception of two boys, all young people involved in our study made extensive use of SNS. These social communication platforms have also become important spaces where young people produce and consume creativity. As such, profile pages aggregate and combine *adapting*, *making* and *creating*. Whereas changing a profile picture or organising and labelling photos are forms of adapting, composing status messages and comments are forms of making and even of creating, as, for example, in the creative use of emoticons and abbreviations.

The gaming practices of Eric, Vince, Diane, Tom, Alex and Justin, as well as the contributions Clare made to Wikipedia, are examples of adapting. In a role-playing game (RPG), as well as in a simulation game, the player is responsible for creating and maintaining one or more characters and for acting its roles. Designing and modifying avatars and spaces, levelling the avatars' skills, managing virtual groups such as a guild, a family or a company are all adaptive activities. These expressions of creativity take place within the predetermined margins of the game software as young people can only chose between what is available. Drawing upon Sanders (2005), gamers do this so that their avatar is better suited to the functional needs of the game. For example, in the RPG *World of Warcraft* a character is specifically designed and skilled in order to play a particular role in the guild when entering a battleground.

While a RPG locks the player into controlled scenarios that have been designed to develop specific competencies and to reach a predetermined goal (for instance, moving to the next level), a simulation game leaves more room for narrative creativity. As such, simulation games provide more possibilities for adapting, for example, avatars, while the game objects are personalised so that they fit the players' (created) personality better. In *The Sims*, Diane not only chose the clothes her family members wore, she also decided the character's traits and parenting style. In *MegaTycoon*, Alex had to decide the goods he wanted his company to produce, the purchasing policy, the staffing policy, etc.

The contributions of Clare to Wikipedia are another example of adapting creativity. As she explained to us, she not only read a lot of articles, but also adapted them and contributed to articles about her family's home country; hence, she made or built something with her hands and mind that did not exist before (see Sanders, 2005). Her creative activities were part of a continuous process of remixing and/or writing over what had come before. Drawing upon Bruns (2008: 3), this form of online creativity corresponds to *produsage*, collaborative content creation and development. Again, the distinction between adapting and making is hard to make, as both processes are part of one practice.

Obviously, this comment also applies to the distinction between making and creating. As soon as young people are engaged in making activities, the most advanced form of creativity, i.e. creating, shows up. Several creative practices listed in the table fit with the notion of productive creativity as they resulted in a concrete outcome. For the occasion of Valentine's Day, Tom compiled a CD with the favourite songs of his girlfriend and other love songs. To do this he looked for songs on his computer and on the Internet, organised them in Windows Media player, copied them to a disc and used a photo taken with his mobile phone for the CD cover. For the eighteenth birthday of a friend, Eve designed a personalised magazine in Microsoft Word and compiled a booklet with real and fictitious stories, an imaginary interview and lots of photos. Catherine, for her part, created a PowerPoint presentation that expressed her love and support for a friend who was ill. With Windows Movie Maker, Eve made a collage of photos taken during a school trip to Italy. The clip's theme was fatigue; the images showed classmates resting or sleeping on the bus, in the hotel, at a historic location, etc. To make the clip more than an animated slide show and to underline the narrative, Eve added the R.E.M. song *Daysleeper*. Last examples are the animated slideshows Clare made of photos of her family and favourite R&B artists with the online application Slide. She shared the results in the community of the same name.

Tom and Alex made several videos with their mobile phone and published them on YouTube. Tom's videos were a loose collection of short clips taken at different places and moments in his personal life: school trips, class hours, lunch breaks, parents' evenings, dinner at home, hanging out with friends in a snack bar, in the park. While Tom's videos did not originate in a clear concept, Alex's videos made up a collection in which each episode had the same concept and narrative structure. Alex directed and filmed the stunts of his friend Peter in four acts: introduction (explaining the stunt), preparation (materially and mentally), performing (doing the stunt) and conclusion (reflecting on the experience). Alex, as well as Tom, did not edit their recordings; they portrayed the action as it took place.

The motivation behind these creative efforts is clear: passionately expressing oneself, using raw materials or recombining existing material in the absence of a predetermined pattern (Sanders, 2005). But at the same time, some of the creations also correspond to "making" as some kind of guidance, e.g., a pattern, a recipe, or notes were used to describe what types of materials to use and how to put them together. Clearly, the guidance young people are relying on when making something through digital technologies is pre-eminently that of the cultural industries, i.e. the music, film and television industry. For example, the gifts Catherine and Eve made, as well as the slides published by Clare, were mash-ups of photos, stories, ideas and images drawn from their own makings and from products of popular culture, respectively the blockbuster film *Shrek*, the popular Flemish youth magazine *Joepie* and photos of R&B musicians. The central concept of Alex's videos was based on popular reality television with shows such as *Jackass*. In other words, the majority of the creative practices we observed in our study can be looked upon as playing with the form and content teenagers find in mass media products. The symbolic meanings they consume are reworked in their own creative practices, as they copy, cut, paste and remix the symbolic forms and content that media industries are producing and distributing.

The remaining creative practices, i.e. the videos by Eric, the drawings by Frank and the game Justin made as a dissertation project, are the most innovating examples of creating. Passion and self-expression are combined with a high level of field experience and an aspiration to innovate. The three aforementioned practices of Eric, Frank and Justin can be categorised as such. Eric is a passionate handyman, who created toys and directed several videos, made with his father's camera. In the movies, Eric held the position of director, cameraman and actor. To him inventing a situation and script, acting a role, and filming and editing the content were part of an intensive creative process, in which special attention was given to the synchronisation of images and sound in Windows Movie Maker. For his part, Frank spent much of his leisure time on drawing. Although he still preferred paper and pencils, the computer screen and mouse in general, and Microsoft Paint in particular, became alternative interfaces for self-expression. Finally, Justin had been passionate about gaming since he was twelve. He not only loved to game, he was also curious about the underlying technology.

In all three cases, the importance of field expertise, acquired through formal and informal learning processes, was clearly demonstrated. During our panel research, Justin was a final year pupil in Information Science. The RPG he created was the innovative outcome of applied knowledge acquired from classes in project development, design and implementation, multimedia, networks, etc. For several years Frank had been following art lessons twice a week. However, not only formal, but also informal learning results in field experience. Eric learned much from his elder brother and his father, who is a journalist. Next to the importance of social environment, these three boys taught us that digital creativity also results from a trial-and-error approach when editing audiovisual content, drawing or programming. As such, expertise is also fuelled by passionate efforts to be satisfied with the end result, in terms of self-expression, esthetical value and technical skills.

Friendship and Humour in Playing with Commercial Texts

When looking at the creative practices of our panel, we noticed that humour, and in particular parody, was a recurring element. Through humour and parody, the teenagers did not only create new meanings in their productions, the integration of funny elements also provided a critical stance on texts or cultural practices. Following Jenkins' (2006) approach, parody can be seen as more than just play for the sake of playing, it is also about redistributing power from the producer, allowing the consumer to contest and critique values, and create his own media (Willett, 2009).

For example, in Alex's videos *Peter eats a moth* and *Peter sniffles salt* some hard-core masculine values were displayed. However, some of his productions also raised questions about masculinity. As the videos *Peter polishes his nails* or *Peter shaves his legs* demonstrated, dominant masculinity was also parodied. These elements also recurred in the fictitious magazine Eve made for the birthday of a friend. On the one hand, Eve valued their relationship characterised by intimacy, closeness and intense collaboration. On the other hand, by making a comic interpretation of the popular girly magazine *Joepie* she also questioned dominant feminine identities.

As these examples show, humour plays a significant role in the consolidation of peer group cultures by offering a space for conveying masculine and feminine identities. The making of parodies of media texts, therefore, works to cement the relationships of teenagers. Texts do not only express teenagers' identities and gives them a common point of reference; they are also a way of making sense of gendered representations.

In friendship having a laugh and to drivel is all part of the game. Friends laugh with all kinds of things, including each other's weaknesses. The friendly relationships represented in the videos of Alex or Tom were also *joking relationships*, to borrow Gluckman's (1963) terminology, in which communication was ambiguous. The jokers were actually making hurtful statements about their friends' appearance, personality or poor performance in the videos, but in a giggly way. It often remained unclear whether what was said was meant seriously. In this kind of mutual teasing or drivelling the friend was implicitly tested on his ability to react. He had to respond to the subdued attack on his personal sense of honour – often on the shared ideas and values related to masculinity – in a humorous way to counter this, otherwise he risked a momentary loss of face.

Confirmation and Distinction: "Come and See Me, Make Me Smile..."

Above, we defined contributions to SNS and forums as interpersonal communication, a form of creativity that is not always recognised by young people as mass self-communication. Nevertheless, these practices provide valuable insight into the social rewards everyday creative communication offers. Our study also shows that throughout the social-communicative applications and personal web pages young people are presenting themselves, and are looking for feedback by others. The responses to their online presence provide teenagers with validation about their *selves*. For example, when Lisa redecorated her bedroom she posted several photographs on the forum she participated in. In the following hours she checked the community for feedback. The same happened when she got herself a new haircut. In a conversation we had afterwards, Lisa admitted that the positive reactions provided confirmation about her image, lifestyle and personality, and made her feel confident about her personal taste. Whereas in the past the bedroom or a new haircut were shown only to some intimate friends who were asked for their opinion, today teenagers angle for an opinion from a much larger social network of online friends. By receiving positive comments their perceived identity and self-esteem increase. The observation that young people find it important to identify with friends through SNS, where they sense their personal and social identity in the interaction with others, shows that the communicative part of creativity is a central occupation in their lives. In online communities teenagers seek for equilibrium between their personal and social identities.

However, young people do not only seek for confirmation, they also try to differentiate themselves from others and as such build their own personality. For example, An published several of her lyrics on her profile page. In conversations we had with her it became clear that in expressing herself online, An aimed to differentiate herself from others by showing personal, original and self-made content. The comments she got valorised this kind of publicity in terms of artistic value, that is, in terms of novelty and authenticity. She was praised for her original lyrics and rewarded by being described as a future artist. Lisa was proud of being in the top ten of people who had posted the most messages, and who had started the most new topics. The user-statistics confirmed Lisa as one of the lead-users. In consequence, she differentiated herself from the majority of members who seldom contributed.

Confirmation and distinction also appeared in the motivation teenagers gave for creating a profile page and in their ideas about SNS. The teenagers made a distinction between on the one hand a MSN space and on the other hand a profile on Netlog, High or Partypeeps. While MSN spaces were considered as *regular*, as they belong directly to the Windows Live Messenger account (used for Instant Messaging), a profile page on Netlog, High or Partypeeps was looked upon as *special*. These profile pages were accompanied with a much stronger sense of belonging to a community, due to the fact that one can build new connections on the basis of shared interests. In other words, having a MSN space involved our participants in a form of normal online participation, while being involved in more niche-like communities, was seen as a way of distinguishing oneself from the regular network (and confirmation in the group). The profile pages on Myspace were an exception. The teenagers with a Myspace account thought they had to become a member of the community in order to be able to listen to the music. In fact, Nicolas, Lisa and Clare were a little embarrassed for having a Myspace page without being artists themselves. They had nothing to contribute artistically, nothing to distinguish themselves from others who were not able to be more than *just fans* either.

Public Rewards of Sensing the Audience

Our observations revealed that forms of self-expression did not necessarily have to be experienced with others; neither did we find that manifestations of creativity were only meaningful to the content creators when an audience was present (Pachucki *et al.*, 2010). This was particularly shown by the drawings by Frank or the videos by Eric, mentioned earlier. Hence, although most of the young people were engaged in various online communication activities, a substantive part of the creative practices was not being shared online. We noticed that the young people in our panel were most creative when they felt motivated primarily by the enjoyment, interest, satisfaction, and challenge of the work itself, rather than by external pressures.

During the conversations two reasons surfaced for keeping content offline. Firstly, the older girls Eve and Julia looked upon their creative expressions as part of their private and intimate interpersonal communication with other friends, mainly girls. In their opinion, making and creating content was closely linked to their inner life. It was about opening up one's heart to and sharing one's deepest feelings with others. Hence, the vulnerability that these girls showed about publishing their own creations online, an aspect Stern (2008, p. 105) also refers to, stemmed from the emotional and interpersonal meaning of their creations.

Besides this, the degree of digital literacy seemed to be another important reason for not publishing online. Two aspects emerged from the interviews: the access to digital technologies and the strategic skills to handle the soft- and hardware for one's own particular use and profit (see Steyaert, 2002; Livingstone, 2004). For example, in explaining why they did not publish their user-made videos some teenagers mentioned a lack of equipment, as they did not have a cable to connect their mobile phone to the computer. Others, in particular the boys from the youngest age group, explained that they did not know how to publish content online, as no one had explained or instructed them how to get started with it. In other words, the reasons for keeping digital creations outside the arena of public exposure and mass reception were divergent, but always rooted in the digital socialization process young people were going through. As we learned from our talks and observations of young people, the role of peers (e.g. in An's case), or parents and siblings (e.g. in Eric's case) seemed to be crucial in understanding how teenagers creatively engage with the digital technologies they have at their disposal.

Building on these findings, we noticed that young people do want to share the outcomes of digital creativity, but not necessarily in the public sphere as enabled through the Internet. This distinction we here tentatively define as *intro-creativity* and *extra-creativity*. With *intro-creativity* we mean creative practices that young people keep for themselves or share only with a relatively small group of people with whom they have strong ties. The aforementioned gifts are examples of intro-creativity, as they were meant for one specific person (girlfriend, best friend). The objects (CD, magazine, slideshow) were created for a specific occasion (Valentine's Day, birthday, illness), and as such symbolised a close relationship (love, friendship). With these gifts the young people expressed and valued an emotional and intimate bond; here the Internet was not considered as an appropriate medium, because it would not value the personal character of the gift. These creative practices also showed that creativity was not driven by the prospect of a mass audience, but by the enjoyment, the interest, and the challenge of the creative process itself. Getting publicly rewarded was in most cases not the primary reason for being creative, nor for sharing the outcome online.

To us *extra-creativity* refers to these creative practices that are publicly shared outside the social network of strong ties. From our panel study we learned that young people used different approaches to conceptualise and construct an anonymous audience for the results of doing, making, adapting and creating. Obviously, the so-called Web 2.0 sites, originally designed to be either user data repositories (such as YouTube) or social networks (like Myspace) (Lin, 2007), are an interesting field for audience researchers to examine how ordinary media users/consumers envision the audience. We learned from our study that when the sense of belonging to a community of shared interests becomes much stronger and when the openness of the SNS goes beyond the social network of actual or physical life (e.g. friends from school, youth movement, summer camps), the awareness of mass audiences, but also of specialist ones increases. This showed, for instance, in the way two boys in our panel, Tom, and to a lesser extent Alex, began to explore the size and range of their audience when they started publishing their own videos on YouTube. Although the young creators still saw their audience in terms of peers, they found it important that their audience no longer solely overlapped with their group of friends and acquaintances. Tom especially hoped for a "fame effect", since he created his account on YouTube to attract people to a forum on gaming he himself had launched. In order to attract and maximise the audience, they explored and applied smart marketing techniques. Firstly, they carefully qualified their videos through well-thought out titles, descriptions and tags, making it clear that the video was a parody of a media text and not a serious imitation of it. At the same time they used two tags for the entertainment-seeking audience that is assumed to be looking for mere *fun* and *foolish* pleasures. However, they also hoped to attract people that were on the Internet looking for other things. So Tom, for instance, tagged his videos as *sex* and *tits*, although this had nothing to do with the actual content of the videos. Again, considering the tags that the boys provided, it shows that they conceptualised their audience as teenage boys, the social category applying to themselves, but at the same time they attempted to target the global market of teenage boys. Although we observed on YouTube that the audience for the boys' videos did not go far beyond the villages where they lived, and that the comments they got were mainly from the producers' immediate circle of friends and acquaintances, they clearly constructed an audience segment in YouTube's vast mass audience and showed a strong awareness about the techniques media industries use to attract consumers.

Within our panel study we dealt with another way of looking at the global audience. One girl in the panel, Clare, contributed actively to Wikipedia by adapting and uploading articles on the home country of her parents. Although she also envisioned a mass audience with worldwide reach when adapting and creating content, the "produsage" paradigm in which she was creative, encouraged her to see the audience as the other people who are co-building and supporting the project of Wikipedia. Hence, Clare's motivation for contributing to Wikipedia was based on reciprocity. Since she learned a lot from the contributions of others, she wanted to give something back to the Wikipedia platform and all the visitors of the online encyclopaedia who are looking for information on the home country of her parents.

Concluding Summary

In this article we explored the social value of online creativity. We have tried to avoid the idea that the only social benefits of a creative usage of technology and the Internet are those that are conventionally educational and goal-oriented. Rather, we have tried to understand how digital creativity enables young people to cope with everyday social life. All participants in our research demonstrated that digital creativity is a way of sustaining friendships, of reproducing values shared with peers, and of challenging dominant values supplied by the cultural industries. Digital creativity is *hard work* (Drotner, 2008), as it is closely connected to expressing one's own identity, so young people take their acts of making, adapting and creating symbolic content seriously. At the same time, digital creativity is also attractive, as it makes it easier to express one's deepest feelings to others, and to show who you *really* are, even if this means playing with form and content one gets from the music and film industry.

The empirical material we have drawn upon taught us that there is a strong relationship between looking for confirmation and distinction on the one hand, and exploring social fields where one can openly play with, explore and validate one's own identity on the other hand. Clearly, identities are still defined in collective terms, but are challenged through a reflective form of identity construction in which daily digital creativity plays an important role. As such, building an identity today is the outcome of an intensified interaction between a wide spectrum of opportunities for self-publication and for social interaction on the Internet. However, despite what a lot of the rhetoric on young people and digital culture suggests, forms of self-expression do not necessarily gain social meaning when they are experienced with a large audience. Young people cherish the privacy and intimacy of their creative explorations and expressions. That is something to be going on with in further investigations of the generation C.

References

Alvesson, M. & Sköldberg, K. (2000). *Reflexive Methodology: New Vistas for Qualitative Research.* London: Sage.

Barker, C. (2008). *Cultural Studies: Theory and Practice* (3rd ed.). London: Sage.

Bauwens, J., Pauwels, C., Lobet-Maris, C., Poullet, Y., & Walrave, M. (2009). *Cyberteens, Cyberrisks, Cybertools: Tieners en ICT, risico's en opportuniteiten.* Ghent: Academia Press.

Beghetto, R.A. & Kaufman, J.C. (2007). Toward a Broader Conception of Creativity: A Case for "Mini-C" Creativity. *Psychology of Aesthetics Creativity and the Arts, 1*(2), 73-79.

Bloustien, G. (2007). "Wigging People Out": Youth Music Practice and Mediated Communities. *Journal of Community & Applied Social Psychology, 17*, 446-462.

Bourdieu, P. (1979). *La distinction. Critique sociale du jugement*. Paris: Les Éditions de Minuit.

Bruns, A. (2005). *Gatewatching: Collaborative Online News Production.* New York: Peter Lang.

Bruns, A. (2008). *Blogs, Wikipedia, Second Life, and Beyond: From Production to Produsage.* New York: Peter Lang.

Buckingham, D. (2003). *Media Education: Literacy, Learning, and Contemporary Culture.* Cambridge: Polity Press.

Buckingham, D. (ed.). (2008). *Youth, Identity, and Digital Media.* Cambridge: MIT Press.

Burgess, J. (2006). Hearing Ordinary Voices: Cultural Studies, Vernacular Creativity and Digital Storytelling. *Continuum: Journal of Media & Cultural Studies*, *20*(2), 201-214.

Burgess, J. (2007). *Vernacular Creativity and New Media.* Ph D thesis. Creative Industries Faculty, Queensland University of Technology.

Crockett, L., Losoff, M., & Petersen, A.C. (1984). Perceptions of the Peer Group and Friendship in Early Adolescence. *Journal of Early Adolescence*, *4*(2), 155-181.

De Block, L. & Buckingham, D. (2007). *Global Children, Global Media: Migration, Media and Childhood.* Basingstoke: Palgrave.

Donath, J. & Boyd, D. (2004). Public Displays of Connection. *BT Technology Journal*, *22*(4), 71-82.

Castells, M. (2009). *Communication Power*. Oxford: Oxford University Press.

de Leeuw, S. & Rydin, I. (2007). Migrant Children's Digital Stories: Identity Formation and Self-Representation Through Media Production. *European Journal of Cultural Studies*, *10*(4), 447-464.

Dowdall, C. (2009). Masters and Critics: Children as Producers of Online Digital Texts. In V. Carrington & M. Robinson (eds.), *Digital Literacies: Social Learning and Classroom Practices* (pp. 43-62). London: Sage.

Drotner, K. (2008). Leisure is Hard Work: Digital Practices and Future Competencies. In D. Buckingham (ed.), *Youth, Identity, and Digital Media* (pp. 167-184). Cambridge: MIT Press.

Ellison, N.B., Steinfield, C., & Lampe, C. (2007). The Benefits of Facebook "Friends": Social Capital and College Students' Use of Social Network Sites. *Journal of Computer-Mediated Communication*, *12*(4), article 1. Retrieved December 2, 2010, from http://jcmc.indiana.edu/vol12/issue4/ellison. html.

Fiske, J. (1990). *Understanding Popular Culture* (2nd ed.). London & New York: Routledge.

Gluckman, M. (1963). *Custom and Conflict in Africa*. Oxford: Blackwell.

Goodman, L., MacCallum-Stewart, E., & Munsell, V. (2009). Spectatorship and Action Research Performance Models. In A. Oddey & C. White (eds.), *Modes of Spectating* (pp. 41-57). Bristol & Chicago: Intellect.

Hammersley, M. (1992). Deconstructing the Qualitative-Quantitative Divide. In J. Brannen (ed.), *Mixing Methods: Qualitative and Quantitative Research* (pp. 39-55). Aldershot: Avebury.

Holloway, S.L. & Valentine, G. (2003). *Cyberkids: Children in the Information Age.* London: Routledge.

Hermes, J. & Janssen, S. (2006). De nieuwe contentmakers. In J. de Haan & C. van 't Hof (eds.), *Jaarboek ict en samenleving 2006. De digitale generatie* (pp. 159-175). Amsterdam: Boom.

Jenkins, H. (2006). *Convergence Culture. Where Old and New Media Collide.* New York: University Press.

Lenhart, A., Madden, M., Rankin Macgill, A., & Smith, A. (2007). *Teens and Social Media.* Washington: Pew Internet & American Life Project. Retrieved December 2, 2010, from http://www.pewInternet.org/~/media//Files/Reports/2007/PIP_Teens_Social_Media_Final.pdf.

Lin, K.-J. (2007). Building Web 2.0. *Computer, 40*(5), 101-102.

Livingstone, S. (2008). Taking Risky Opportunities in Youthful Content Creation: Teenagers' Use of Social Networking Sites for Intimacy, Privacy and Self-expression. *New Media & Society, 10*(3), 393-411.

Livingstone, S., Haddon, L., Görzig, A., & Ólafsson, K. (2010). *Risks and Safety on the Internet: The Perspective of European Children. Initial Findings.* London: LSE, EU Kids Online.

Livingstone, S. (2004). What is Media Literacy? *Intermedia, 32*(3), 18-20.

Loveless, A.M. (2006). *Literature Review in Creativity, New Technologies and Learning.* Report 4. Future Lab Series. Retrieved December 17, 2010, from http://www.futurelab.org.uk/resources/documents/lit_reviews/Creativity_Review.pdf.

McDermott, M. (2008). Youth Media Production in the Context of a Community-based Media Arts Project. In D. Buckingham (ed.), *Youth, Identity, and Digital Media* (pp. 185-206). Cambridge: MIT Press.

Mediappro (2006). *Mediappro: A European Research Project: The Appropriation of New Media by Youth.* Final Report. Brussels: EC.

Mesch, G. & Talmud, I. (2010). *Wired Youth: The Social World of Adolescence in the Information Age.* New York: Routledge.

Niesyto, H., Buckingham, D., & Fisherkeller, J. (2003). VideoCulture: Crossing Borders with Young People's Video Productions. *Television & New Media, 4*(4), 461-482.

Ofcom (2006). *Media Literacy Audit: Report on Media Literacy Amongst Children.* Retrieved December 17, 2010, from http://stakeholders.ofcom.org.uk/market-data-research/media-literacy/medlitpub/medlitpubrss/children.

Pachucki, M.A., Lena, J.C., & Tepper, S.J. (2010). Creative Narratives Among College Students: Sociability and Everyday Creativity. *The Sociological Quarterly*, 51, 122-149.

Papacharissi, Z. (2002). The Presentation of Self in Virtual Life: Characteristics of Personal Home Pages. *Journalism and Mass Communication Quarterly*, *79*(3), 643-660.

Pruulmann-Vengerfeldt, P., Kalmus V., & Runnel, P. (2008). Creating Content or Creating Hype: Practices of Online Content Creation and Consumption in Estonia. *Cyberpsychology: Journal of Psychosocial Research on Cyberspace*, *2*(1), Article 1. Retrieved October 20, 2010, from http://cyber psychology.eu/view.php?cisloclanku=2008060202&article=1.

Sanders, L. (2005, March). *Information, Inspiration and Co-creation.* Paper presented at the 6th International Conference of the European Academy of Design, Bremen, Germany.

Schlesinger, P. (2007). Creativity: From Discourse to Doctrine? *Screen*, *48*(3), 377-387.

Sefton-Green, J. & Buckingham, D. (1998). Digital Visions: Children's "Creative" Use of Multimedia Technologies. In J. Sefton-Green (ed.), *Digital Diversions: Youth Culture in the Age of Multimedia* (pp. 62-83). London: Routledge.

Sefton-Green, J. & Soep, E. (2007). Creative Media Cultures: Making and Learning Beyond the School. In L. Bresler (ed.), *International Handbook of Research in Arts Education. Part 2* (pp. 835-854). Dordrecht: Springer.

Sima, Y. & Pugsley, P.C. (2010). The Rise of a "Me Culture" in Postsocialist China. *The International Communication Gazette*, *72*(3), 287-306.

Stern, S. (2008). Producing Sites, Exploring Identities: Youth Online Authorship. In D. Buckingham (ed.), *Youth, Identity, and Digital Media* (pp. 95-117). Cambridge: MIT Press.

Steyaert, J. (2002). Inequality and the Digital Divide: Myths and Realities. In S. Hick & J. McNutt (eds.), *Advocacy, Activism and the Internet* (pp. 199-211). Chicago: Lyceum Press.

Tacchi, J. (2004). Researching Creative Applications of New Information and Communication Technologies. *International Journal of Cultural Studies*, *7*(1), 91-103.

Walsh, D. (2004). Doing Ethnography. In C. Seale (ed.), *Researching Society and Culture* (2nd ed.) (pp. 225-237). London: Sage.

Willett, R. (2009). Consumption, Production and Online Identities: Amateur Spoofs on YouTube. In R. Willett, M. Robinson, & J. Marsh (eds.), *Play, Creativity and Digital Cultures* (pp. 54-67). New York: Routledge.

Willis, P. (1990). *Common Culture: Symbolic Work at Play in the Everyday Cultures of the Young.* Milton Keynes: Open University Press.

Youngsters and Online Political Participation

A Story in Need of Qualification

Peter THIJSSEN

Introduction

Most observers agree that today's citizens, and particularly the younger generations, are highly critical and demanding of their political representatives (Norris, 1999; Pharr & Putnam, 2000). This is apparent from, among other things, their volatile voting behaviour, their weak party identification, and ultimately even their political disaffection (Dalton & Wattenberg, 2001). As a consequence, political representatives are becoming increasingly uncertain and sensitive to signals from their electoral base. In social psychological terms, one could say that most representatives are extreme self-monitors, i.e. individuals who are acutely aware of the public perception of their own behaviour (Snyder, 1974). Hence the impact of public political participation on those representatives has arguably never been as great as it is today, all the more as those representatives are more accessible than ever before due to the introduction of online participatory tools. In this context, it is especially relevant to consider possible biases in political participation, particularly in respect of online variants.

The outlines of offline political participation are, after all, well established (Verba, Schlozman, & Brady, 1995; Blais, 2000; Rosenstone & Hansen, 2003), and the so-called *digital divide* is now also empirically attested (e.g. Norris, 2001; Sacco, 2002; Shane, 2004; van Dijk, 2005). It was initially hoped that the opportunities offered by online participation would overcome the biases in traditional participatory tools, but this optimism has since given way to pessimism (Castells, 1996; Gibson, Lusoli, & Ward, 2005). Most researchers now tend towards the *reinforcement perspective*, according to which new online participatory tools aggravate existing biases due to the fact that resources that are relevant in offline political participation, such as time, knowledge and money, also codetermine people's degree of online political participation.

However, many feel there is one variable that does not fit into this pattern, namely age (Delli Carpini, 2000; Best & Krueger, 2005; Di

Gennaro & Dutton, 2006; Dalton, 2008; Saglie & Vabo, 2009). Whereas youngsters on the whole score worse in respect of traditional offline political participation, according to these authors this might not be the case for new online participatory tools. The lower political participation rate among youngsters is generally attributed to a negative lifecycle effect: in the course of their lifetimes, individuals accept a growing number of social commitments, such as parenthood and homeownership. As a consequence, they have growing interests to protect and hence are more inclined to participate in politics. This effect is further enhanced by the fact that individuals become gradually accustomed to participating (Plutzer, 2002; Franklin, 2004). In online political participation, however, the negative lifecycle effect is neutralised by a generation effect that is due to the specific socialisation context. First and foremost, youngsters today are the first generation to have grown up in the computer age. As a consequence, they possess Internet skills, a unique resource that substantially reduces the cost of online political participation. Second, today's younger generations are generally better educated and they are encouraged much more greatly to call into question hierarchical authority (Inglehart, 1999). According to Dalton, one should therefore look beyond the reduced electoral participation of the young and acknowledge that they participate through other, more varied, channels. They opt resolutely for more direct, individualised and issue-specific methods. Dalton has argued that low-cost, low-information types of participation (Verba *et al.*, 1995), which are usually performed on the basis of a sense of citizen duty or habit (Blais, 2000; Zukin, Keeter, Andolina, Jenkins, & Delli Carpini, 2006), are rejected by youngsters in favour of participatory tools that allow them to challenge and influence the authorities more directly. Moreover, youngsters want greater freedom in the timing of their participation: they are prepared to participate, but only when it suits them. In this respect, the Internet is the perfect participatory medium for the younger generations. Instead of flocking to the polling booth, they prefer to contact politicians or civil servants directly, from the comfort of their own desks. As Dalton puts it, "the Web is the medium of the young – and often an unknown world to their grandparents" (Dalton, 2008, p. 66).

However, there are also authors who question the impact of such generation effects (e.g. Putnam, 2000; Wattenberg, 2002; 2008). In this context, Wattenberg (2002, p. 90-91; 2008, p. 28-29) and Putnam (2000, p. 177) refer respectively to the phenomena of *narrowcasting* and *cyber-balkanisation*. Unlike traditional broadcasters, the new (Internet) media focus on a specific topic and a specific target group. Hence, in today's Balkanised media landscape, one can shut oneself off from political information. Hence, it is by no means self-evident that youngsters will use their Internet skills for political purposes. The aforementioned authors are therefore sceptical about the validity of available empirical data: "Some of the allegedly greater democracy is based more on hope and hype than on careful research" (Putnam, 2000, p. 173).

Although the above assertion testifies to a degree of defeatism, there is clearly a need for more qualified research into age effects in online political participation rates, particularly considering ongoing demographic evolutions and the fact that the present financial and economic crisis will most likely compel politicians in the Western industrialised world to make policy choices with tangible generation and age-specific consequences. In sum, if the stakes are substantial and if decision-making by politicians is indeed led by the perceived desires of the public, then it would be problematic for young stakeholders to let their voices be heard less emphatically.

The Need for More Focused Research

It seems odd that such prominent political scientists disagree so fundamentally over a basic fact that appears to be easily verifiable. In our view, the explanation for the ambivalence of the available research findings in relation to age-related variation in online political participation is attributable to the often over-simplified research approach that ignores a number of important nuances in respect of conceptualisation, theory and data.

First, current political research takes little or no account of the diversity that exists in online political participation (Bimber, 1999; Polat, 2005). After all, one should distinguish appropriately between the consultation of a political website, the sending of an email to a political representative and the participation in a debate on a political forum, because each of these activities is based on different skills and, even more importantly, they tap into different needs. Moreover, even if it emerges that the various forms of participation correlate substantially and constitute a one-dimensional scale, the question remains whether this also holds for different age groups.

Second, the theoretical starting point often lies in resources theory, which focuses mostly on the cost of participation while largely ignoring its benefits (Verba *et al.*, 1995; Campbell, 2003). The benefits of participation are considered to play an important role in the explanation of lifecycle effects, and we see no reason why they should not be relevant to generation effects, too. For that matter, one should take into account here benefits originating in participation as such, as well as those stemming from participation as a social activity (Blais, 2000). The latter benefits, which are based more on value rationality than on instrumental rationality, are often neglected, but according to Inglehart (1997) they are nonetheless more important to youngsters today. For the young generations of today, the satisfaction derived from being involved in the participatory process may be at least as important as generating policy impact.

Third, current research is generally based on fairly standardised survey data involving the reported behaviour of national samples. A first problem with this approach is that national samples are often not relevant because online populations are generally confined by national boundaries. Obviously, if one wishes to obtain a truly representative picture of online political participation by various age groups, the general and anonymous context of the Internet can be a serious drawback (Cammaerts & Van Audenhove, 2005). However, by relying on national samples one it is very hard to draw conclusions with regard to the representativeness of the findings. A second problem is that most survey research relies on reported online political behaviour. Respondents are asked whether they have recently made use of online tools for political participation. Nonetheless, methodological research has shown very clearly that a survey response is often subject to considerable bias, particularly in the context of social desirability (Verba *et al.*, 1995, p. 106; Schuman & Presser, 1996). It is, for example, not unimaginable that younger respondents, who tend to regard online activity as trendy, will exhibit a tendency to exaggerate their online political activity. The third problem is related to the use of fairly standardised questions, such as "Have you last year participated in an online forum?". Obviously, these type of questions make life easier when doing comparative research, but at the same time they hamper in-depth insights in the specific participatory benefits. In this respect carefully designed case-studies focusing on actual participants are very valuable.

Fourth, an additional explanation for the persistence of the controversy regarding the political behaviour of youngsters is that the issue is often framed in generational terms. Although political participation research now has a well-established tradition, the necessary longitudinal datasets to definitively settle the controversy are not available. After all, this would require a comparison between the political participation of present young cohorts and that of older cohorts at the same age. This is however complicated by the fact that the dependent variables, i.e. the observed forms of political participation, have themselves changed. The traditional batteries of questions concerning political participation primarily gauged the degree of participation in collective political action, such as elections, campaign activity and local community work. Today, citizens possess many more possibilities for making their voices heard and these new individualised forms of participation are not comparable to the traditional collective and institutionalised channels of action. More specifically, the possibilities for participation offered by the Internet are too recent to allow adequate generation-based comparisons. As long as the controversy is framed in generational terms it is to settle it. Therefore, in the present contribution, we shall not endeavour to draw conclusions about generation effects. Instead, we restrict ourselves to age-related differences in online political participation taking due account of the previously made comments. This way, we base ourselves on population data concerning effective participation in two distinct forms of online political participation. Moreover, with a view to explaining the observed age-related differences, we shall also focus on age-specific benefits that (may) stem from participation.

Research Questions and Context

As we have previously mentioned, it is our express intention to gain insight into revealed rather than reported online political participation by various age groups. To this end, we consider two Belgian case studies offering a fairly adequate picture of effective online political activity on the part of people of different ages. Case studies are inevitably vulnerable to the argument that they may be unique or atypical but as we argued before the value added through knowledge of the population and the participatory context constitutes a significant advantage over the diffuse nature of much of the international comparative survey data. Moreover, the Belgian context is potentially instructive because of its relatively large age-specific interests. Not only does Belgium have a strongly developed social security system, it also has a pensions scheme that is based in part on repartition. This implies that pensions are partly funded by the active population, which makes the notion of intergenerational transfers particularly tangible. In other words, in the Flemish context, the various age groups may be assumed to have good reason to make their voices heard to their political representatives.

In each of the two case studies we aim to answer the following two research questions: "Do youngsters excel in political participation?" and "If they do, then is this to be understood as being due to the singularity of the participatory context?". In the first case study, we look at the age structure of the citizens that report issues to local government by means of an Internet tool. To this end, we rely on a population database of citizens having made use of the local Internet hotline in the district of Deurne (Antwerp, Belgium) in the period 2004-2007. The district of Deurne has a population of 70,000. It is one of several districts making up the City of Antwerp, with a total population of 400,000. We explicitly focus on the age structure of three subgroups. Firstly, local citizens that have reported an issue on one or more occasions by email. Secondly, local citizens that have filed offline problem reports. Thirdly those citizens not having reported problems at all. In the second case study, we concentrate on the age structure of the membership group of an independent online political forum called *Politics.be*. Towards the end of 2006, we organised a web survey of members of this forum. As we were also granted access to objective log data, we were able to distinguish between *posters*, i.e. those who posted at least once in the course of 2006, and *lurkers*, i.e. those who merely made passive use of the forum.

Although both cases concern forms of online political participation, there are significant differences between the two. Online reporting follows a top-down logic, whereby the authorities initiate the medium. Activity on a political forum, on the other hand, is essentially horizontal in nature. Assuming that youngsters are more attracted to non-hierarchic horizontal participation while older citizens are more familiar with traditional reporting tools, it is not unthinkable that the age structure of the groups of users of the two forms of online political participation are different (Amnå, 2010). Moreover, the sending of an email is an activity that fits into an instrumental rationality, whereas the discursive context of participation in a political forum ties in more with a value rationality.

The initial focus in both case studies is on the complete age structure of the group of participants, but special attention shall be paid to the young. However, a methodological note is in place here. Many of the disputes over age-related discrepancies concern the delineation of the age groups. More specifically, the question arises where the boundary lies between old and young. It may be specified on the basis of divergent series of generation-related criteria, though admittedly these are always slightly arbitrary. Like most political scientists, Wattenberg (2002) relies on the legal retirement age to define the category of the old. In the United States, as in Belgium, this age is set at 65 years. Clearly, though, it is a legal threshold that rarely corresponds with individuals' actual retirement age. Nor does it coincide with any other meaningful lifecourse transition. As regards the upper threshold for the group of youngsters, this is commonly set at the age of adulthood, which in most countries is 18 years. Again, though, this boundary seems increasingly artificial, particularly in view of an extended adolescence. Moreover, in the context of voting participation research, this would lead to the absurd conclusion that there are, strictly speaking, no young voters.

It is therefore proposed that age boundaries should be approached more circumspectly. Where we do make use of a categorisation, the older group shall be delineated by means of two thresholds. The first is the age of 58, when the average Belgian worker exits the labour market, but is still willing and able (to continue) to engage socially. The second threshold is that of 70 years, when the average Belgian and/or their partners begin to experience health issues that tend to compromise their continued social engagement. As regards the upper threshold of youth, this is expanded from 18 years to 34 years.

Case Study 1: Local Problem Reporting by E-mail

The Data

The dataset for the first case study consists in all reports by email or a report form submitted to the district authorities of Deurne in the years 2004, 2005, 2006 and 2007 (Thijssen, 2009). The website and the report form contained the following introductory statement: "Use this document for free reporting of problems and suggestions relating to the local street environment or urban services". In total, 4303 reports were filed. The annual number of reports for respectively 2004, 2005, 2006 and 2007 was 1,068, 1,054, 1,082 and 1,099. The research focus was first and foremost on individuals who had filed a report. After elimination of institutional reporters, who had participated in a professional capacity, 2,451 "unique" individuals were retained who had reported at least once in the course of the four-year timeframe.

It was our intention to describe this group of 2,451 persons as accurately as possible. To this end, the individuals concerned were linked with the population register on 1 August 2008. The overall success rate of this exercise was 79.6% or 1,950 cases. Reasons for a mismatch included moving house, incomplete or inaccurate personal details, and death. As a basis for comparison was required, the 1950 retained reporters were, on the one hand, contrasted with the remaining population of Deurne. To this end, and with assistance from the Department for City Observation of the City of Antwerp, a complementary dataset was compiled. This dataset related to the 69,036 residents of Deurne who had not reported during the period surveyed. Finally, the group of 1,950 reporters was divided into 721 reporters by email and 1,229 reporters by means of the traditional paper report form.

The Age Differences

Figure 1 compares the age distribution of the group of 721 individuals who reported at least once by email with that of the 69,315 residents of Deurne who filed no reports in the 2004-2007 period. For the time being, age is regarded here as an uncategorised continuous variable. The pattern that emerges is clearly curvilinear. In the group aged approximately 35 to 70, there are disproportionately many residents who reported a problem by email. Under age 35 and above age 70, there are comparatively fewer reporters by email. Both the average and the median age of the reporter by email is 51, which indicates that the distribution is fairly symmetrical. Immediately, Dalton's assumption that the Internet is an exclusive terrain of the young is called into question somewhat. It may well be the case that youngsters spend more time at the computer, but apparently not to report on local policy issues. Furthermore, the data suggest that familiarity with the Internet from a young age is not a requirement to use this medium. On the face of it, the group reporting local problems by email consists mainly in residents of middle age. But also among the senior citizens, there are quite a few individuals who seem attracted to the comfort and speed offered by the digital information highway. Wattenberg would appear to be right in asserting that, also in relatively new and electronically mediated forms of participation, the younger group is somewhat outnumbered by the older generations, though it should be pointed out that this observation does not hold for the oldest group (the over 70s). In other words, if the age variable is to be categorised, it would seem worthwhile also to distinguish between the over 50s and the over 70s.

Figure 1: Reports by e-mail

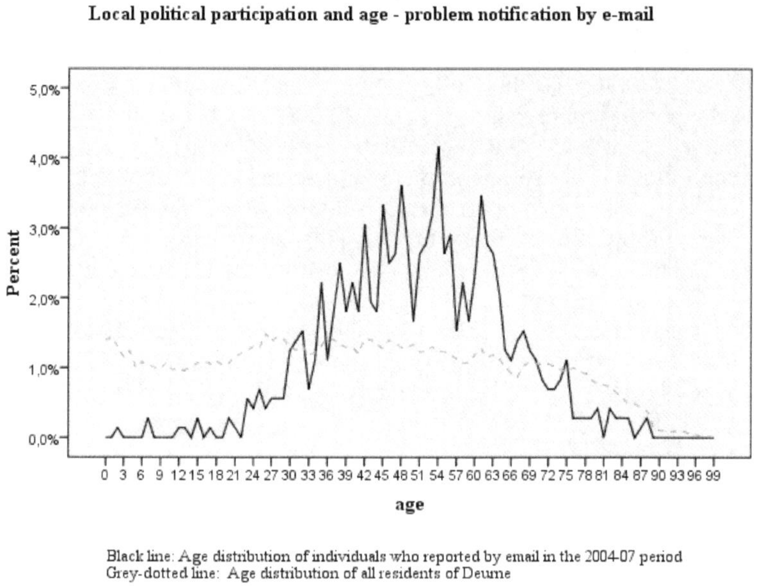

Figure 2 compares the age distribution for the 1,229 reporters by paper form with that for the 69,315 residents of Deurne who did not report in the 2004-2007 period. Again, the percentage distribution of reporters is curvilineair, or parabolic to be more precise. This time, however, the top is slightly broader and flatter. From around age 40 to age 85, there are consistently disproportionately many residents who reported issues by traditional offline means (paper report form, face to face). Once again, the young are less inclined to report. In fact, this time, the number of reporters is low up to age 40. The average age of the non-electronic reporter is 59, which is 8 years older than in the case of electronic reporting. A noticeably large proportion of the over-70s report problems to the local authorities. Apparently this group opts exclusively for traditional communication channels.

Figure 2: Reports by paper form

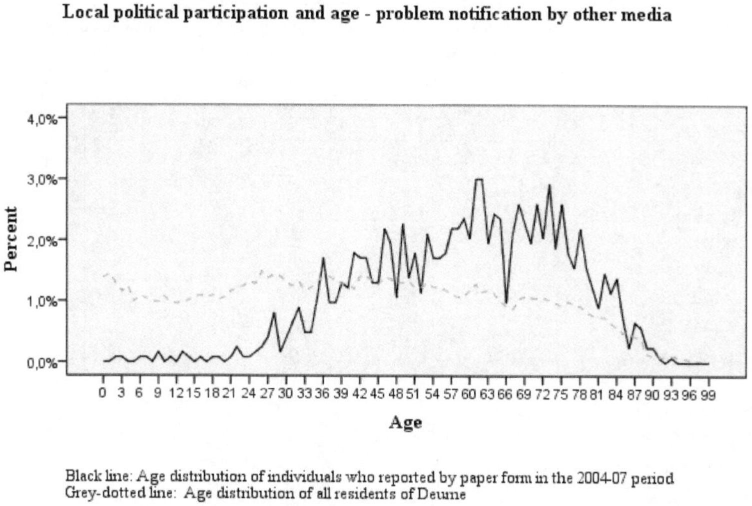

So in both offline and online reporting, the observed relationship is parabolic. The results of the logistic regression in Table 1 illustrate that this effect persists in a multivariate context.[1] A parabolic relationship is after all represented mathematically by means of a second-degree polynomial of the form $Y= \beta_0 + \beta_1 X + \beta_2 X^2$ and the parameter estimates of the logistic regressions indicate that both age and age squared have a significant effect on the likelihood of an individual reporting. The sign of age is positive, suggesting that older citizens are more likely to report. At the same time, however, the sign of age squared is negative, indicating that the likelihood of reporting declines among the oldest respondents. Equally interestingly, the effects persist after controlling for the period of residence. As people reside longer in a particular location, they tend to show more involvement with local policymaking.[2] We do how-

---

1 These analyses are restricted to the adults in the research population (18 years or older).

2 The residence period is, most probably, usable as a proxy for homeownership. Certainly in Belgium, people who have lived in the same location for long period of

ever observe that the age effects are more pronounced among reporters by email. Hence, the results of the logistic regressions lend credence to the reinforcement perspective, particularly as the differences in terms of gender and ethnicity are also greater among those reporting their grievances online. Males and native residents tend to report more, especially when they file their reports by email.

Table 1: Who reports to the local authorities?

	Report by email versus no report		Report by offline medium versus no report		Report by email versus complaint by offline medium	
	B	S.E.	B	S.E.	B	S.E.
Zage	5.472***	.426	3.592***	.296	1.931**	.586
$Zage^2$	-4.810***	.362	-2.451***	.223	-2.402***	.477
ZResidence period	.217***	.048	.201***	.035	.068	.064
Gender						
Female	.000		.000		.000	
Male	.659***	.080	.483***	.060	.301*	.106
Ethnicity						
Foreign	.000		.000		.000	
Non-foreign	-1.364***	.161	-.598***	.107	-.517*	.194
Family composition						
Single, no children	.000		.000		.000	.000
Single with children	-.078	.195	.095	.134	-.286	.242
Couple with children	.508***	.114	.058	.091	.269°	.152
Couple, no children	.421***	.110	.008	.072	.253°	.139
Other	.070	.368	-.494	.341	.587	.515
Constant	-5.816***	.136	-4.895***	.099	-.927***	.176
n	712		1,218		712	
N	55,634		56,140		1,930	
% correctly classified	.987		.958		.642	
Nagelkerke Pseudo R^2	.077		.058		.135	

Note: Logistic regression analysis, Dependent variable: Issuing a complaint: Yes=1 and No=0; *** $p < 0.0001$; ** $p < 0.001$; * $p < 0.01$; ° $p < 0.10$.

time tend to own their dwelling. Unfortunately, we possess no data on the extent of homeownership, nor for that matter on individuals' educational level, as this information is not included in the population register.

The Individual Benefits of Local Problem Reporting

Socio-economic data show that people in their thirties are generally better educated than individuals from older generations. In principle, the younger generations should therefore find it slightly easier to report a problem. However, as Andrea Campbell (2003) rightly emphasises in her book *How Policies Make Citizens*, one needs to take into consideration not only the costs but also the benefits of participation (*stakes*). She asserts, for example, that social security policy involves US senior citizens more closely with politics and that it is conducive to their participation. As many of these citizens are dependent upon allowances, they have more to lose than the younger generations.

A similar reasoning may be applied to local issues. Firstly, the elderly, particularly in Belgium, generally own the dwelling in which they live, whereas this is far less the case among those in their thirties. As home values are influenced by the quality of the location, the elderly stand to benefit more from problem reporting about the local street environment. Furthermore, the frequency of house moves is much lower among homeowners than among renters. It is plausible that a longer residence period in a particular neighbourhood may also lead to greater local involvement. Finally, the older generations are usually no longer professionally active, so that they may be assumed to be less outgoing, and consequently more sensitive to local problems. In sum, the younger generations most likely participate less because they feel less involved in the neighbourhood where they live, e.g. because they tend to regard it more as a temporary home or a place to sleep. As Meyrowitz (1986) puts it, the younger generations have less of a sense of place. To the extent that local participation and sense of place are connected, this would appear to be a potentially strong age effect that is mitigated only marginally by the nature of the participatory medium.

In order to gain insight into the age-related concerns of local participants, Table 2 provides an overview of the kind of topics covered in problem reporting.[3] As could be expected on grounds of the above argumentation, it emerges that the elderly are relatively more concerned about problems relating to the local street environment, while youngsters are more preoccupied with mobility issues, such as free parking passes for local residents, even though in absolute terms the elderly by far outreport the young in relation to the latter type of issues, too.

---

[3] For individuals who filed more than one report, it was the topic of the first report that was taken into consideration. In 26.8% of reports, it was impossible to ascertain the issue reported on, as these were filed by telephone.

Table 2: Type of issues reported by age

Reporter 2004-2007 Type of Issue	Age -35	35-55	56-70	70+	Total
Security	2	2	1	1	6
	2.0%	.4%	.2%	.3%	.4%
Greenery	13	77	72	45	207
	13.0%	13.5%	15.7%	15.2%	14.5%
Environment and waste	14	78	63	28	183
	14.0%	13.7%	13.7%	9.4%	12.8%
Housing	5	11	4	3	23
	5.0%	1.9%	.9%	1.0%	1.6%
Street environment	35	254	219	163	671
	35.0%	44.5%	47.6%	54.9%	47.0%
Community problems	8	29	25	9	71
	8.0%	5.1%	5.4%	3.0%	5.0%
Public services	2	22	15	10	49
	2.0%	3.9%	3.3%	3.4%	3.4%
Mobility	21	98	61	38	218
	21.0%	17.2%	13.3%	12.8%	15.3%
Total	100	571	460	297	1428

Case Study 2: Participation in an Online Political Forum

Data

In this second case study, we focus on the audience participating in an online political forum called *Politics.be* (Thijssen, 2008). *Politics.be* (forum.Politics.be) is an independent forum for debate on domestic and international politics, moderated by volunteers. Despite being a Dutch language forum visited mainly by Flemish users, *Big Boards* (www.big-boards.com) ranks it among the busiest of its kind in the world. Today, it has a membership of around 16,000 and the number of messages posted has passed the four million mark. Between Friday 25 August and Saturday 14 October 2006, a web survey was sent to the addresses of the then 7,433 registered members of *Politics.be*. In all, 1,329 persons responded. The request to participate in the survey came from researchers, but was disseminated by the moderators of *Politics.be*. The initial email of 25 August was followed by reminders on 6 September and 4 October.

So the web survey reached 17.9% of the forum's registered members. However, this figure is slightly misleading, as the longevity of email addresses tends to be short. Given that the online forum *Politics.be* was launched in 2002, it is reasonable to assume that a substantial proportion of the originally registered email addresses were no longer valid at the time of the survey. Moreover, it is also reasonable to assume that some individuals registered several times on the forum, but responded to the survey only once.

The overall picture becomes clearer after comparison of the subjective survey data with the objective log data, which we also had at our disposal. Of the 1,115 members who posted at least one message during the period from 1 January 2006 to 14 October 2006, no fewer than 438 participated in the survey. In other words, the response rate among posters was 39.3%. If we focus on the 369 most intensive posters during that same period, each of whom posted over 100 messages, the response rate rises further to 50.1%. The 1,330 respondents accounted for 188,042 messages during the aforementioned period, which amounts to 51.1% of the total number of messages (367,549) posted within that timeframe.

Age Differences

Figure 3 below represents the age distribution of the population of 1,329 registered members who participated in our web survey. Unlike in the previous case study, it would appear to confirm Dalton's assumption that youngsters do not participate less in politics, but merely differently. Participation in online political debates would seem to be a case in point: the graph exhibits two prominent peaks, representing young adults aged between 18 and 25. Subsequently, it declines more or less continuously. Strikingly, though, the decline is rather slow between 25 and 60 years, after which point it accelerates significantly.

Figure 3: Age structure of forum membership

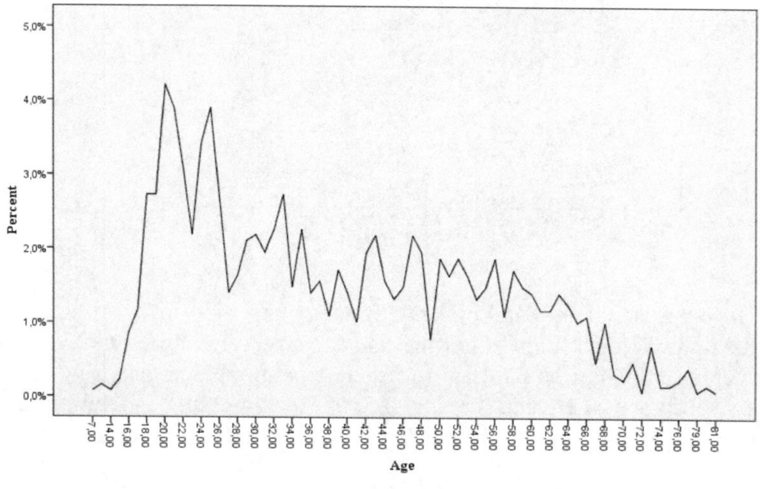

The question arises whether a similar trend is discernible if we look separately at the 563 lurkers (42.4%) and the 766 posters (57.6%). Lurkers are registered users who follow the debates but do not participate actively in them. The percentage of lurkers is strikingly high, as there is no need to register in order to follow the debates. Most probably, many of these lurkers once had the intention of posting, but never actually got round to doing so.

Figure 4 shows that the preponderance of youngsters is much stronger among the posters (black line) than among the lurkers (grey line). The peak representing the young adults, for example, is much more pronounced among the posters than among the lurkers. It is noticeable that the proportion of respondents who merely followed the debates passively is similar for those in their thirties and those in their fifties. Once past age 60, however, the proportion of forum participants declines rapidly among both posters and lurkers. Clearly, then, seniors are not dominantly present on this online political forum, a finding which contrasts starkly with observations in the previously discussed case study. However, as has been pointed out, there are substantial differences between the two types of electronically mediated participation studied.

Figure 4: Age structure of the groups of posters and lurkers

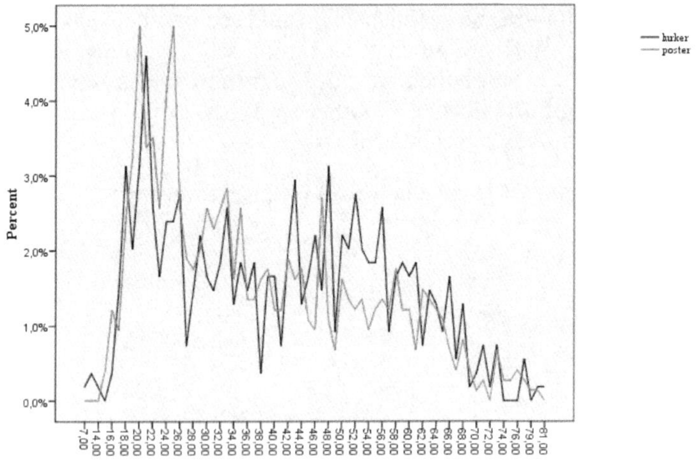

In contrast to the parabolic relationship observed in the case study focusing on a local Internet hotline, the age structure of the group of participants in this online political forum appears to be a negative function. The differences between the passive lurkers and the active posters, however, come to the fore most clearly among the younger respondents.

Proportionally, there are many more posters among the young respondents. In this sense, it comes as no surprise that the multivariate testing represented in Table 3 below revealed a statistically significant reciprocal effect of age.

Table 3: Who posts on the online political forum?

	Poster on online forum versus lurker	
	B	S.E.
Zageinv (1/Age)	.238***	.062
Gender		
Female	.000	
Male	.250	.168
Educational level		
Primary	-.760°	.317
Lower secondary	-.563*	.190
Higher secondary	-.292°	.130
Higher	.000	
Constant	.314°	.171
N	719	
N	1,242	
% correctly classified	.585	
Nagelkerke Pseudo R^2	.031	

Note: Logistic regression analysis, Dependent variable: Poster=1 & Lurker=0; *** p < 0.0001; ** p <0.001; * p < 0.01; ° p < 0.10.

The Individual Costs and Benefits of Forum Participation

Unlike with electronic reporting of a problem, which is an instance of vertical communication between public and government, debate between forum members is an example of horizontal communication. Furthermore, problem reporting serves a concrete policy purpose, whereas the purpose of participation in a political debate is primarily symbolic. Hence, in problem reporting, the benefit to be gained is essentially extrinsic and users' motivation is driven by an instrumental rationality, while in the case of forum participation, the benefits are essentially intrinsic and users tend to be driven by value rationality.

Table 4 below provides an overview of the principal reasons for the various age groups to participate in the political forum.[4] For the lurkers, the forum serves first and foremost as a correlation and information tool. On the one hand, they wish to gain insight into what other people think. On the other, they want to stay informed about politics. This is not so surprising, as these functions tie in quite nicely with the passive attitude of this type of forum user. Yet, even here we see a remarkable difference in terms of age. Whereas the youngest generation of lurkers is focused firstly on the correlation function (36.1%) and only in the second instance on the information function (20.3%), the preferences among the older generation of lurkers is the reverse. Among the young seniors (56-70 years), for example, 24.1% gives precedence to the correlation function compared to 37.9% to the information function.

It is striking how the correlation function is also important to the posters. In the active age group (35-55 year) and among the young seniors (56-70 years), this is by far the most commonly cited reason for using the forum. Once again, though, the group of the youngsters (under 35 years) diverges slightly from the rest of the respondents. The modal motive among respondents from the youngest generation is deliberative in nature: they use the forum to discuss politics and society. Somewhat remarkably, this deliberative motive is less prominent among the older generations. They are apparently less interested in the pattern of argument and counterargument that underlies any debate. The older respondents see the forum first as an information channel and, additionally, as a medium that allows them to express their opinions freely. To them, interaction is not a goal in itself. For that matter, this observation ties in with the traditional socialisation models, where it is assumed that primary political socialisation occurs mainly during childhood and adolescence. Given the phenomenon of prolonged adolescence, we may assume that many of the respondents belonging to the youngest age group are still in the process of forming a political identity. Unlike the older groups, they are therefore interested more in confronting their own ideas with those of others. The interest of the older forum users, on the other hand, is focused mainly on the finding raw information, or possibly a sounding board, but not so much an interlocutor. A possible explanation for this observation is that they are also able to satisfy those needs via the more traditional media, such as readers' letters to newspapers. In this sense, they are perhaps less dependent upon online forums. Seen against this backdrop, it would not seem coincidental that young adults, aged between 18 and 25, are particularly active on political forums.

[4] It concerns the answer to the following question: "Which of the following reasons do you feel is the most important for participating in the forum?"

Table 4: Reasons for forum participation by age

		Age				Total
		-35	35-55	56-70	70+	
lurker	Insight into the views of others	57	49	21	5	132
		36.1%	29.9%	24.1%	38.5%	31,3%
	Keeping up to date with politics	32	49	33	5	119
		20.3%	29.9%	37.9%	38.5%	28,2%
	Forming a personal opinion	8	4	4	0	16
		5.1%	2.4%	4.6%	.0%	3,8%
	As a pastime	5	9	6	0	20
		3.2%	5.5%	6.9%	.0%	4,7%
	Representing an organisation or party	3	1	2	0	6
		1.9%	.6%	2.3%	.0%	1,4%
	Making one's voice heard	15	13	4	0	32
		9.5%	7.9%	4.6%	.0%	7,6%
	Getting to know people	1	0	0	0	1
		.6%	.0%	.0%	.0%	,2%
	Discussing politics and society	21	19	2	1	43
		13.3%	11.6%	2.3%	7.7%	10,2%
	Being able to express one's opinion freely	13	19	12	2	46
		8.2%	11.6%	13.8%	15.4%	10,9%
	Convincing others	3	1	3	0	7
		1.9%	.6%	3.4%	.0%	1,7%
Total		158	164	87	13	422
		100,0%	100.0%	100.0%	100.0%	100.0%
poster	Insight into the views of others	80	69	35	3	187
		26.0%	33.7%	35.0%	17.6%	29,7%
	Keeping up to date with politics	41	30	14	6	91
		13.3%	14.6%	14.0%	35.3%	14,4%
	Forming an opinion	19	8	3	0	30
		6.2%	3.9%	3.0%	.0%	4,8%
	As a pastime	18	6	5	1	30
		5.8%	2.9%	5.0%	5.9%	4,8%
	Representing an organisation or party	16	7	1	0	24
		5.2%	3.4%	1.0%	.0%	3,8%
	Making one's voice heard	23	28	13	2	66
		7.5%	13.7%	13.0%	11.8%	10,5%
	Getting to know people	1	0	0	0	1
		.3%	.0%	.0%	.0%	,2%
	Discussing politics and society	81	26	11	1	119
		26.3%	12.7%	11.0%	5.9%	18,9%
	Being able to express one's opinion freely	21	29	15	3	68
		6.8%	14.1%	15.0%	17.6%	10,8%
	Convincing others	8	2	3	1	14
		2.6%	1.0%	3.0%	5.9%	2,2%
Total		308	205	100	17	630
		100.0%	100.0%	100.0%	100.0%	100.0%

Be that as it may, participating in an online forum continues to be an activity reserved for a limited elite, because, although the benefits for youngsters resulting from participation in such a medium probably outweigh the costs, the latter should not be overlooked. We are concerned here with a form of political participation that, to most individuals, is relatively knowledge-intensive and time-consuming, and that is therefore founded more on intrinsic than on extrinsic motivations. Hence, this type of political participation may be said to be geared mainly to the niche group of the higher educated. This certainly holds for the posters, of whom 50.2% has a higher education degree, whereas the corresponding proportion among the lurkers is only 42.6%.

The Costs and Benefits for the Various Age Groups

As has previously been mentioned, the younger members of *Politics.be* are attracted to the forum mainly because it offers an opportunity to engage in political debate. The older generations are either interested readers or in search of a sounding board. A substantial proportion of the young forum members is, after all, still developing a political identity. In this context, it is interesting to consider the ideological profile of their interlocutors, particularly those from the older age groups. To this end, we again used the scores of a subjective self-identification scale from 0 to 10. The scores were subsequently subsumed into three categories: 0 to 4 and 7 to 10 are qualified as respectively left-leaning and right-leaning, while the 5 and 6 are considered to be centre. In Table 5 below, we see that the users of *Politics.be* tend, on the whole, to be right-leaning. The respective proportions of right-leaning individuals among the lurkers and the posters are respectively 46.2% and 44.9%. Although this aspect was not taken into consideration in the first case study, it is interesting to compare with the political profile of the population of Deurne: the proportion of right-leaning reporters and non-reporters was respectively 30.0% and 24.3%, which is a substantially smaller percentage than among the members of *Politics.be.*

It is noticeable that there is a rather substantial difference between the ideological profiles of the youngest age group and the older age group. The youngsters are significantly more left-leaning than the elderly. Among young lurkers and posters, the proportions of left-leaning forum members is respectively 35.8% and 33.8%, whereas the corresponding proportions in the young senior group are 15.4% and 12.1%. In view of the previously reported observation that the younger members are intent on engaging in a debate, and therefore need an adversary, this is not so surprising. Nonetheless, it is hard to predict what effect such a polarised debate shall have on political attitudes. Authors such as Sunstein (2001) seem to assume that this can only lead to extremism. However, additional research on this matter is called for, as the aforementioned authors base their opinions primarily on experimental studies with rather strict starting points. Be that as it may, it would seem safe to assume that the youngsters will be more easily influenced by the debates than the elderly. The analysis of the motivations for forum participation did after all indicate that the older respondents, the vast majority of whom are in fact young seniors, are not so much seekers of as respondents to debate. They enjoy being able to "*express their opinions freely*", which generally entails publicising their right-wing ideas. It is therefore possible that the older forum members, despite being less well represented than the youngsters, have a significant impact. However, this matter requires more in-depth research, for instance based on content analyses of the forum discussions.

Table 5: Age groups and position on the political spectrum

			Age				**Total**
			-35	**35-55**	**56-70**	**70+**	
lurker	Left-right	Left	62	50	14	1	127
			35.8%	26.0%	15.4%	7.1%	27.0%
		Centre	43	51	28	4	126
			24.9%	26.6%	30.8%	28.6%	26.8%
		Right	68	91	49	9	217
			39.3%	47.4%	53.8%	64.3%	46.2%
		Total	173	192	91	14	470
			100.0%	100.0%	100.0%	100.0%	100.0%
poster	Left-right	Left	113	59	13	4	189
			33.8%	27.7%	12.1%	23.5%	28.2%
		Centre	93	52	34	2	181
			27.8%	24.4%	31.8%	11.8%	27.0%
		Right	128	102	60	11	301
			38.3%	47.9%	56.1%	64.7%	44.9%
		Total	334	213	107	17	671
			100.0%	100.0%	100.0%	100.0%	100.0%

All in all, it is quite surprising that so many older persons use a political participation medium that ties in closely with chat culture, especially since many of them appear to approach *Politics.be* as an information channel. The explanation is most likely closely connected with the (extreme) right-wing profile of many members of the forum. After all, the individuals concerned are generally very politically motivated and keen to use any opportunity that presents itself to disseminate their ideas, particularly as options to do so via the more traditional channels are limited. Many of them seem to regard *Politics.be* as a kind of reserve where they are able to ventilate their opinions unrestrictedly. In this sense, we are concerned here with a form of political participation that attracts those who feel they are insufficiently heard in the more traditional participatory channels.

Conclusion

While authors such as Dalton (2002; 2008) maintain that the decline in traditional electoral participation by youngsters is compensated for by their increased involvement in new forms of political participation, others, like Wattenberg (2008), claim that youngsters are also outdone by the older generations in the latter participatory tools. Dalton is particularly optimistic about the opportunities for participation offered by the Internet. Wattenberg, too, emphasises the impact of new information technologies such as the Internet, but his outlook is a lot more negative. He argues that, as the new media are conducive to narrowcasting, whereby the focus is concentrated on a specific topic, they allow youngsters to shut themselves off from complex political information. According to Wattenburg, youngsters do use the new information technology for emailing and chatting, but not about politics. In sum, the question at hand could be rephrased in terms of Dalton's optimistic view versus Wattenberg's pessimistic assessment.

Initially, the evidence gathered seemed to tend more strongly towards Wattenberg's pessimistic hypothesis. After all, the first case study indicated that the older respondents have certainly discovered the possibilities of participating online in local politics. In fact, the 56 to 70-year-olds make greater use of online reporting to their local authorities than the youngest age group (under 35 years). Perhaps this is due to the fact that youngsters are less concerned with the local street environment for reason that they are not homeowners and therefore tend to regard the places where they live as temporary residences. Be that as it may, the online medium also creates opportunities for youngsters to raise issues that are of concern to them, e.g. local sports or recreational infrastructure.

The second case study, however, paints an entirely different picture. We found that youngsters (under 35 years) are strikingly well represented among the users of the online political forum *Politics.be*. Apparently, then, a distinction needs to be made between different forms of online political participation. Although the first case, like the second, concerned an instance of online communication, clearly the rules of play are not very different from those governing traditional offline political participation. The role of the playmaker, so to speak, is primarily fulfilled by the authorities. In this sense, all that is required from the older generations is a limited capacity for adaptation. Matters are quite different in our second case study, which relates to activities on an online political forum. Here, the older users are, at best, forced into the role of observers. This finding coincides with the observation that the informal and decentralised discursive logic of the forum deviates quite fundamentally from the formalised and government-initiated consultation logic of the Internet hotline.

Still, all things considered, quite a substantial group of older persons appears to be active on the online political forum. Dalton's assertion that the web is the exclusive realm of the younger generation does not hold. What is more, it emerges that youngsters and older persons have different reasons to become forum members. The younger participants are interested mainly in political debate, whereas the older members use the forum firstly as an information source and secondly as a sounding board. Hence, it is no coincidence that the proportion of lurkers is greater among the older group. Nonetheless, it is not unthinkable that the older posters have a greater influence on forum activity than their relatively small number would suggest, although further research is needed in order to be able to draw robust conclusions regarding the age-related impact of forum participation. However, while the older generations seem reluctant to access online political forums, most political representatives appear equally reluctant to draw lessons from the political glossolalia on such platforms. To the extent that politicians are more inclined to acknowledge participants in traditional Internet applications, such as direct contact by email, there is a danger that they are made more aware of the concerns of older citizens than those of the younger generations. Therefore, in addition to exploring how different age groups use divergent possibilities for online political participation, it would be worthwhile to conduct research into the attitudes of politicians *vis-à-vis* the various online participatory tools.

References

Amnå, E. (2010). *New Forms of Citizen Participation: Normative Implications.* Baden-Baden: Nomos.

Best, S.J. & Krueger, B.S. (2005). Analyzing the Representativeness of Internet Political Participation. *Political Behavior*, *27*(2), 183-216.

Bimber, B. (1999). The Internet and Citizen Communication with Government: Does the Medium Matter? *Political Communication*, *16*(4), 409-428.

Blais, A. (2000). *To Vote or Not to Vote. The Merits and Limits of Rational Choice Theory*. Pittsburgh: The University of Pittsburgh Press.

Cammaerts, B. & Van Audenhove, L. (2005). Online Political Debate, Unbounded Citizenship, and the Problematic Nature of a Transnational Public Sphere. *Political Communication*, *22*(2), 179-196.

Campbell, A.L. (2003). *How Policies Make Citizens. Senior Political Activism and the American Welfare State*. Princeton: Princeton University Press.

Castells, M. (1996). *The Rise of the Network Society*. Oxford: Blackwell.

Dalton, R.J. (2002). *Citizen Politics: Public Opinion and Political Parties in Advanced Industrial Democracies*. Chatham: Chatham House Publishers.

Dalton, R.J. (2008). *The Good Citizen. How a Younger Generation Is Reshaping American Politics.* Washington DC: CQPress.

Dalton, R.J. & Wattenberg, M.P. (2002). *Parties Without Partisans. Political Change in Advanced Industrial Democracies.* Oxford: Oxford University Press.

Delli Carpini, M. (2000). Gen.com: Youth, Civic Engagement, and the New Information Environment. *Political Communication, 17*(4), 341-349.

Di Gennaro, C. & Dutton, W. (2006). The Internet and the Public: Online and Offline Participation in the United Kingdom. *Parliamentary Affairs, 59*(2), 299-313.

Franklin, M.N. (2004). *Voter Turnout and the Dynamics of Electoral Competition in Established Democracies Since 1945.* Cambridge: Cambridge University Press.

Gibson, R., Lusoli, W., & Ward, S. (2005). Online Participation in the UK: Testing a "Contextualised" Model of Internet Effects. *British Journal of Politics and International Relations*, *7*(4), 561-583.

Goerres, A. (2007). Why Are Older People More Likely to Vote? The Impact of Ageing on Electoral Turnout in Europe. *British Journal of Politics and International Relations*, *9*(1), 90-121.

Hibbing, J.R. & Theiss-Morse, E. (2002). *Stealth Democracy. Americans' Beliefs About How Government Should Work.* Cambridge: Cambridge University Press.

Inglehart, R. (1990). *Culture Shift in Advanced Industrial Society.* Princeton: Princeton University Press.

Inglehart, R. (1997). *Modernization and Postmodernization. Cultural, Economical, and Political Change in 43 Societies.* Princeton: Princeton University Press.

Manin, B. (1997). *The Principles of Representative Government.* Cambridge: Cambridge University Press.

Meyrowitz, J. (1986). *No Sense of Place. The Impact of Electronic Media on Social Behavior.* Oxford: Oxford University Press.

Norris, P. (1999). *Critical Citizens. Global Support for Democratic Governance.* Oxford: Oxford University Press.

Norris, P. (2001). *Digital Divide. Civic Engagement, Information Poverty, and the Internet Worldwide.* Cambridge: Cambridge University Press.

Pettersen, P.A. & Rose, L.E. (1996). Participation in Local Politics in Norway: Some Do, Some Don't, Some Will, Some Won't. *Political Behavior*, *18*(1), 51-97.

Pharr, S.J. & Putnam, R.D. (2000). *Disaffected Democracies. What's Troubling the Trilateral Countries?* Princeton: Princeton University Press.

Plutzer, E. (2002). Becoming a Habitual Voter: Inertia, Resources, and Growth in Young Adulthood. *American Political Science Review, 96*(1), 41-56.

Polat, R.K. (2005). The Internet and Political Participation. Exploring the Exploratory Links. *European Journal of Communication, 20*(4), 435-459.

Rosenstone, S.J. & Hansen, J.M. (2003). *Mobilization, Participation, and Democracy in America*. New York: Longman.

Sacco, D. (2002). *Cybering Democracy. Public Space and the Internet.* Minneapolis: University of Minnesota Press.

Shabe, P.M. (2004). *Democracy Online. The Prospects for Political Renewal Through the Internet*. New York/London: Routledge.

Saglie, J. & Vabo, S.I. (2009). Size and e-Democracy: Online Participation in Norwegian Local Politics. *Scandinavian Political studies, 32*(4), 382-401.

Schuman H. & Presser, S. (1996). *Questions & Answers in Attitude Surveys. Experiments on Question Form, Wording, and Context.* London: Sage.

Snyder, M. (1974). Self-monitoring of Expressive Behavior. *Journal for Personality and Social Psychology, 30*(4), 526-537.

Sunstein, C.S. (2001). *Republic.com*. Princeton: Princeton University Press.

Thijssen, P. (2008). Het is stil waar het nooit waait. Online politieke fora als bakermat van publieke opinie. *Res Publica, 50*(2), 169-178.

Thijssen, P. (2009). Wie klaagt waar? Waar klaagt wie? *Bestuurswetenschappen, 63*(4), 47-64.

Van Dijck, J.A.G.M. (2005). *The Deepening Divide. Inequality in the Information Society.* London: Sage.

Verba, S., Schlozman, K.L., & Brady, H.E. (1995). *Voice and Equality.* Cambridge: Harvard University Press.

Wattenberg, M.P. (2002). *Where Have All the Voters Gone?* Cambridge: Harvard University Press.

Wattenberg, M.P. (2008). *Is Voting for Young People?* New York: Pearson Longman.

Zukin, C., Keeter, S., Andolina, M., Jenkins, K., & Delli Carpini, M.X. (2006). *A New Engagement? Political Participation, Civic Life, and the Changing American Citizen.* New York: Oxford University Press.

Children and Online News

A Suboptimal Relationship; Quantitative and Qualitative Research in Flanders

Rozane DE COCK[1]

Youngsters and News: A Lack of Interest

On the basis of the ideal of a society consisting of well-informed citizens, reference is often made to the importance of the news as a crucial source of information and knowledge. Nevertheless, previous research shows that young people have a lack of interest in traditional serious news media (Betzel, Lauf, Negenborn, & Vosselman Bosch, 2008; Costera Meijer, 2006; Raeymaeckers, 2002; Patterson, 2008). Scholars and media professionals have recorded a decline in readership among the whole population with the drop in readership especially notable among young people (Raeymaeckers, 2002). Disappointing audience ratings of newscasts aimed especially at young people such as Zoom and Jam (Vandebosch, Dhoest, & Van den Bulck, 2009) also make it clear that the audiovisual approach via target group programmes does not always result in the desired effect.

What is even more disturbing is that young people do not automatically evolve from non-news consumers into news consumers simply by getting older, as news researchers and media professionals have often presumed. Recently, research has pointed out that there is no such thing as an automatically performing age effect but that one should rather speak of a cohort effect with respect to the consumption of news in relation to age: a whole layer of peers within the population pyramid is not moving up from non-news consumer to news consumer (Costera Meijer, 2006).

Furthermore, young people in effect appear to be less informed about important public issues and news topics compared to previous generations (Evard, 1996). Therefore, the socially relevant function of news as

1 The author is very grateful to the students of the second year bachelor in communication sciences enrolled in a methodology class during the spring of 2008 for their help with the quantitative data collection. Extra thanks goes to Marjolein Put, master in communication sciences, who conducted the qualitative interviews as part of her masters thesis.

a supplier of information and general knowledge, something which is indispensable in order to create a well-informed citizenry that is able to participate in democracy, is liable to be lost. Previous studies indicate that boys tend to watch the television news more often than girls do and that older children watch it more frequently than younger pupils (Smith & Wilson, 2002; Van den Bulck, 2006). The same gender difference reappears when we look at newspaper reading: boys read newspapers more often and for a longer time than girls do (Raeymaekers, 2004). The co-orientation role of the parents also became apparent. The reading frequency of the parents correlated with the reading frequency of the adolescents and the impact of the father's newspaper consumption pattern was shown to be especially influential (Raeymaekers, 2004).

From Adolescents to Children

The above-mentioned news consumption research is aimed specifically at young people, an age group that is defined and demarcated in a variety of ways: 16-18-year-olds, 15-25-year-olds, or pupils of secondary school (Costera Meijer, 2006; Raeymaeckers, 2004; Van den Bulck, 2006). But as far as children are concerned, there is a void in news consumption research. Seen from a developmental perspective, this lacuna in research is hard to understand. According to the cognitive theory of development, the transition from the concrete operational stage to the formal operational stage (in which children develop abstract thought and learn to derive consequences from events) is situated as early as age 11 (Piaget). The structures or thinking schemes that are formed at this age form the basic units of knowledge (De Wit, Van der Veer, & Slot, 2001).

When we think of children between 10 and 12 years old and their news consumption, attention goes mostly to television news. According to the research of Roe (2006) this is precisely the age at which children watch the most television (cognitively more developed children even more so than their peers). In Flanders, the number of young viewers increases steadily between 18 and 21 hours. During that time slot, a vast majority of 4-11-year-olds watches along with their parents (Heene, Vossen, Van Damme, & Buyse, 1999). Moreover, it is precisely in this time slot that the adult version of the main newscast of the public network VRT and the commercial network VTM are broadcast. Foreign research (Drew & Reeves, 1980; Buijzen, Walma Van der Molen, & Sondij, 2007) shows that children watch the adult news more often and at a younger age than is generally thought. Even if children do not select the news by themselves, and although the adult version is not explicitly meant for them, this programme genre has an effect on them (Buijzen, Van der Molen, & Sondij, 2007; Cantor & Nathanson, 1996; Walma Van der Molen, Valkenburg, & Peeters, 2002). Research has shown that people are highly dependent on television as far as their knowledge of news events is concerned.

Transition to Adolescence: Need for Control and the Internet

When children make the transition towards adolescence their needs change, their capacity to understand more complex media increases, and they get more access to and control over their leisure time (Roe, 2006). The changes in the life of a young adolescent become apparent in his/her media use. General research on television viewing behaviour of children between 10 and 12 years old indicates a shift toward more adult-oriented media programme preferences (Valkenburg, 2008) as children look for media over which they have more control. The news makes up a typical media genre that is linked explicitly to adulthood and interest in society (politics, economy, culture etc.).

However, at this particular age children gradually experience less satisfaction from television as a result of which they lose interest in it and focus their attention on other media (Roe, 2006). If this line of reasoning also holds true for television news, it is crucial to take a closer look at the possible transition toward use of news on other (new) media among this age group. Special attention should go to the role the Internet plays in the satisfaction of the desire for information/news and the *need for cognition* (Sparks, 2006), that is, thinking through the consequences and reasons for news broadcasts. Pessimists blame the lack of interest in news among young people on the lack of motivation in processing cognitively demanding information, the loss of social capital and a general absence of interest in the world outside their immediate environment. A more optimistic view of the motivation of young people attributes the reason for their lack of interest in news to the content and the format of the mainstream news itself: news in newspapers and on television has no connection with the social world of young people and as a result they give up on this type of news format. Changes in news format, presentation style and content (more local news, clearer and simpler use of language, a more flashy approach, and so on) can, according to this view, increase the attraction of news bulletins (Baym, 2007). For example, online news is possibly more able to meet these criteria, since it is constantly available, gets permanently updated and is situated on specific news sites; news in this form is served and packaged in the form of a quick snack rather than an indigestible meal (Ahlers, 2006; Costera Meijer, 2006). These specific features of online news may offer children greater control over the medium, in contrast to the limited control a user has when watching the news on television. They feel more in control when they surf the Internet and behave as more active and selective consumers.

The Internet as Source of News: Displacement Effect?

When young adolescents swap classic news media for the Internet as a source of news, a displacement effect might occur. However, Costera Meijer (2006) warns of a news paradox: though standard format news is thought of by many young people as dull and uninteresting to watch, infotainment and other formats, may be appealing but are not really news programmes in their opinion.

In addition, we should be cautious of adopting an overly optimistic technological perspective which states that new media function as complete and only answer to the imbalance of knowledge in our society. We can make three critical comments concerning this perspective. First, we may wonder whether online news sites operated by mainstream media and news logs really do reach a young public, as this form of news service is not yet fully developed (Ngyen, 2008). Online news is on the rise, but three-quarter of all online news users say they came across news quite by accident. Young people especially stress this "coincidence aspect" (Pew Research Center, 2008). Secondly, it remains to be seen whether children and adolescents are explicitly aware of the potential of the Internet as a news source and whether searching for news on the net is a purposeful and deliberate act for them. The Internet is primarily used to keep contact, to search for social contact and to entertain oneself (Gross, 2004; Lenhart, Madden, Rankin Macgill, & Smith, 2007). In the third place, mere access to the Internet is no guarantee that the knowledge gap, digital or not, will narrow (Broos, 2006; Donoso, 2007; Ngyen, 2008; Roe, 2006). In contrast to classic media, online news does not follow one particular S-diffusion curve, one that would determine what percentage of the population has already entered the stage of early adopters or is a late-coming majority (Rogers, 1995), but different S-curves of different online news applications in which faster and better software and hardware are at work at the same time. In this way the skills and access of the child/youngster to the very latest media devices play a decisive role in whether it will be aware of new developments or falls behind (Ngyen, 2008).

Method and Data Collection

A standardised paper survey comprehensible to children was distributed among a representative sample of primary schools in Flanders (age category: fifth and sixth year of primary school, third grade). We contacted schools beforehand in order to obtain their cooperation. While they were taking the survey, the children were always welcome to ask questions of the assistants present in the classroom in order to clarify potential difficulties. In total, 1047 children completed the survey during the spring of 2008. 52% of the respondents were pupils from the fifth year, 48% were in the sixth year. 50.8% of the sample were boys, 49.2% were girls.

In our qualitative research, we conducted 15 in-depth interviews with Flemish children aged 10 and 11 (fifth year of primary school). We selected eight boys and seven girls. The participants (respondents) were interviewed at home without any other person present so participants could speak freely and feel secure at the same time. On average, the interviews took half an hour. During the interview, we used a semi-structured interview guide that focused on the news consumption of the participants, their motives for using news, preferences in topic and news media (ranking TV news, radio news, newspapers and the Internet). The interviews were conducted during the spring of 2009. Every in-depth interview was taped, fully transcribed, and analysed thematically in order to answer our central research question.

Research Questions

If we want young people to develop into well-informed citizens with broad interests, we may have to stimulate their appetite for news consumption earlier and in a different way than has generally been thought up till now. Specific features of online news (interactivity, speed, permanent updates, personal menu, links to further information, etc.) may offer children greater control over this medium than someone has when watching the television news (for instance, no control over item sequence or item selection). These advantages of online news over print media or other audiovisual news media may play a crucial role in the process of stimulating children's interest in news as they make the transition to adolescence. This first of all requires a current state analysis of news consumption among children in Flanders and the media they prefer to use. It is also the main purpose of our study.

Based on the literature and the gaps shown in it, we can formulate the following research questions:

RQ 1A: To what extent are children of the fifth and sixth grade in primary school in Flanders news consumers?

RQ 1B: Do children of this particular age group prefer children's television news or a news site developed for youngsters?

RQ 1C: Do socio-demographic characteristics (gender, age, SES of the parents) impact on the news consumption of children of the fifth and sixth grade in primary school in Flanders?

RQ 2: Does the displacement effect occur when we consider news consumption among children? In other words, is the use of the Internet as news medium already well established among children and does this medium function as a substitute or does this young generation primarily rely on traditional news media?

Results

Children as Consumers of News

Our survey results indicate that traditional media still win the news battle, especially television news, which has many viewers among children of 11-12 years old. The television news for children on the Flemish public channel *Ketnet*: *Karrewiet*, is well known by our respondents: 97 percent of the questioned pupils were aware of this daily programme. *Karrewiet* is watched by 23% of the respondents on a daily basis and by 29% several times a week (in total 52%). Surprisingly, television news for adults is even more popular among children than the children's news. No less than 28% of them watch it every day and 32% of the children watch the adult news several times a week (in total 60%). Half of the sample listens to radio news from at least several times a week to almost every day. Far less attention is paid to the Internet. 37% of the respondents answered affirmatively when asked whether they surfed the Internet (in general) when they were looking for something concerning the news. Our results further indicate that despite efforts of the newspaper industry and policy makers such as the *Newspaper in the classroom* project (*Kranten in de klas*) the printed version of newspapers is hardly read by children. Only 8% say they read a newspaper every day, 15.7% several times a week and 25% of the sample never reads a paper at all.

News Site or Children's Television News

We asked our respondents to indicate on a Likert scale whether they would prefer a well-made television news programme for young people rather than surfing to a news site especially developed for youngsters. The majority of the children (53.7%) said they preferred the television news broadcast, while 46.4% chose the website. This result indicates the importance of television news for children but also shows that a future, well-composed website which is especially adapted for them would find favour in children's eyes. In answer to another question, 45.3% of the respondents claimed they would definitely visit a news site that would be especially designed for the young.

Figure 1: Preference of news format of children of the fifth and sixth year of primary school

	Strongly disagree (%)	Disagree (%)	Agree (%)	Strongly agree (%)	(N)
I would prefer a well-made television news programme for young people rather than surfing to a news site especially designed for youngsters	19.6	26.7	35	18.7	1,029
I would definitely visit a news site specially made for youngsters.	21.3	33.5	32.2	13.1	1,034

Socio-Demographics and News Consumption of Children

Boys and girls do not differ when we focus on newspaper reading ($t(1030)= -.20$; $p >.05$) nor for surfing the Internet for news ($t(1032)= -.57$; $p > .05$). When we look at the other news media, differences do occur. Girls watch children's news more often on television ($M = 4.25$; $SD = 1.76$) than boys do ($M = 3.44$; $SD = 2.17$) ($t(998)= -6.6$; $p < .001$). Listening to the radio news is also more popular among female pupils (M 3.83= $SD = 2.11$) than among their male counterparts ($M = 3.5$; $SD = 2.37$) ($t(1028)= -2.36$; $p < .05$). When we look at the adult version of the television news, the opposite pattern becomes apparent. Boys ($M = 4.44$; $SD = 1.82$) watch the adult television news more often than girls ($M = 4.12$; $SD = 1.74$) ($t(1031)= 2.89$; $p < .01$).

Although correlation coefficients are small, age does correlate positively with surfing the net when searching for news ($r= .07$, $p < .05$), that is, older respondents tend to use this medium more often than younger children. Watching *Karrewiet* (children's television news) correlates negatively with age ($r= -.09$, $p < .01$). Younger children watch this news programme, specially made for children, more than older respondents. Children of the fifth and sixth year of primary school do not differ significantly as far as watching the adult television news, listening to radio news or reading newspapers are concerned.

The socio-economic status of the parents, conceptualised as their educational level measured separately for mother and father, also shows small positive correlations with news consumption per particular medium. The higher the educational level of the parents, the more often children surf the net for news or read newspapers. These correlations are significant for the mother's educational level (spearman's rho= .09 and .07, $p < .05$) as well as for the father's (spearman's rho= .07, $p < .05$).

No significant correlations were found between the educational levels of the parents and use of television news (whether adult or children's versions) and radio news.

Displacement Effect of Online News?

During the in-depth interviews, we let our participants rank news media (TV, radio, the Internet and newspapers) while balancing the pros and cons in order to find out which medium was the most appropriate for news in their opinion and why. Due to a technical failure in one taped interview making this question incomprehensible it brings the total amount of participants for this part to 14. 10 out of 14 participants rank television news first. Only one interviewee put television news last. Not only did the majority of the children think of television as the best news medium, 13 participants also claimed to watch it sometimes or even often. Television newscasts such as *Jam*, *Zoom* and *Karrewiet* are included. 5 out of these 13 said they even watched television news on a daily basis. Only 2 children watched it seldom or never. Images are the main asset explaining the popularity of television news among children. Images make the news more clear and it takes less effort to understand the news compared to reading the newspaper. Furthermore, everyone has the possibility of following television news: all interviewed participants had a television set at home, and no extra effort is needed as is the case in buying a newspaper. Some participants also valued the broad explanation of events given in television news.

INTERVIEWER: Why do you think television news is the best?

THOMAS: Well, [the tv set] is in the living room, you can easily reach it, just turn it on, whatever you want.

INTERVIEWER: Which one is the best in your opinion: television news or radio news?

JOZEFIEN: Television news because that is far more elaborate and then you understand more of it.

A large part of our participants thought of this elaborateness as a demerit: they believed television news was too long which made it dull. Lack of flexibility was also reported as a negative element: this medium has no options to pick and determine the order of topics which sometimes results in long waiting times until the desired item comes along.

The newspaper appears to be a more appropriate news medium than radio according to the interviewed children but is less popular than television news. Half of the participants put newspapers in second place in the rankings. The remaining half put newspapers in the first, third and last place respectively. Eleven participants said they read newspapers. Three participants read a newspaper on a daily basis, the other eight once in a while. Three participants never or seldom read newspapers.

Most participants only looked at the front page and skimmed through the paper. They often only read titles and only when they noticed something interesting, they would read the rest. Some children only picked out comics and games. The time they spend reading the paper varied from two minutes to a quarter of an hour. One participant stated he/she read the newspaper every day for half an hour.

One of the positive characteristics our participants mentioned was the flexibility of this medium: one can take it anywhere at any time. Freedom of choice in topics was another advantage highly appreciated by children. In contrast with television or radio news, they could choose the order of the topics and were able to skip uninteresting sections. One of the interviewees stressed the large amount of information as a plus for newspapers.

INTERVIEWER: Why do you think the newspaper is best?

DORIEN: Well, er, because, they can put more in it and you can choose what you want.

INTERVIEWER: Tv comes last?

ELIEN: Yes.

INTERVIEWER: Why?

ELIEN: It takes too long and in a newspaper you can skip what you do not want to read.

The biggest disadvantage of newspapers is the extra effort it requires to consult this medium in comparison with television or radio and this effort forms a barrier for children. One of the participants said he did not read the newspaper because he simply did not like reading in general.

Radio scored far less well as a news medium than television. In the rankings, radio news mostly ended up in third or fourth and last place. Only one participant preferred radio to television because these newscasts managed to put events in a nutshell, briefly and quickly without wasting time. In addition, radio news was updated every hour as a result of which more up-to-the-minute reports were aired. Furthermore, radio news could be heard anywhere, for instance, when their parents drove them to school or to leisure activities, children listened along with them. Besides the advantages of radio news, the lack of images was the biggest disadvantage of this news medium. It made this type of news more dull, children said. They often considered the radio news as a background sound while they were doing something else. Some participants stated they did not understand a lot of the radio news: it was too difficult for them and it went too fast for the information to sink in and be remembered.

The in-depth interviews show that opinions of the Internet as a news medium differ among children. Eight participants put the Internet last in the rankings. Three participants placed this medium in the second place and only one said it should be ranked first. Nevertheless, we can deduce from the interviews that our participants use the Internet intensely. When we asked them how often they used the Internet, six children said they did this on a daily basis, eight participants answered: "a couple of times a week", and only one child stated it "rarely or never" surfed the Internet. However, the possibilities new media offer for consulting news appear to be rather under used by our young participants. None of our interviewees said they consulted news sites often. Six ended up on news sites from time to time while one participant had used it only once. Six other children knew news sites did exist and were aware of the possibility of reading an online newspaper or of watching the (televised) news as an online version but none of them used it themselves. Two participants were not even aware of the existence of news sites.

> INTERVIEWER: Do you sometimes check the web sites of the news, for instance the site of Karrewiet?
>
> JOZEFIEN: Yes
>
> INTERVIEWER: And newspaper sites?
>
> JOZEFIEN T.: Yes, sometimes. First, I type something in Google and when I do not find anything there, then I think: "I'll have a look at the television news channels [sites]" and if I do not find anything there, then I sometimes go and check newspapers [sites]", but most of the time, I find everything I want by searching Google.

Those children who used news sites thought of it as a convenient medium, for instance when they wanted to know more about a specific news event or when they had missed a broadcast of the television news (for children). They often used a news site because it was the default home page set by one of their family members on the family's computer. The participants always relied on the same news site because they were unaware of, or unfamiliar with, any other news sites, which, as a result, were evaluated as too complicated and too inconveniently arranged. Children mainly used news sites in order to gather information for school. One of our participants said they used news sites because of the competitions that often run on these sites and he was eager to take a shot at one of the prizes. Although children said they loved images (as mentioned earlier when dealing with the motives behind the preference for television news), none of our participants, except one child, watched video images or audiovisual news items on news sites. When children used these news sites, they only turned to the text.

> INTERVIEWER: Do you sometimes surf to the web site of the television news?
>
> JOZEFIEN B.: Most of the times that is Karrewiet.
>
> INTERVIEWER: And what do you do then?

JOZEFIEN B.: When I have missed a news broadcast, I go and check what it was about.

INTERVIEWER: Do you watch the whole television news broadcast on the Internet?

JOZEFIEN B.: No, they write it down there.

Our participants were not aware of the interactive and hypertextual advantages of online news and they did not make use of tools such as online discussion platforms or online audiovisual news items. Only one interviewee said he watched video footage from time to time on the news site of the public channel. Nevertheless, he was not in favour of this feature owing to "environmental" circumstances:

INTERVIEWER: Why do you rank the Internet last?

FERDNAND: Because you can not hear it very well because of the noise and stuff.

INTERVIEWER: You mean because of the others in the house?

FERDNAND: Yes.

Furthermore, other disadvantages of online news were mentioned. One participant doubted the quality of news sources on the Internet because "anyone can publish anything on the Internet". The same child stressed that it was less pleasant to read the news on a computer screen and she preferred a paper version.

INTERVIEWER: Why the Internet last?

STANSKE: Because, well, it is all junk. Well, maybe it consists of the same [content as newspapers] but it is not fun to read it.

INTERVIEWER: So, you prefer reading things in a newspaper rather than on the Internet?

STANSKE: Yes, you can turn the pages.

Conclusion and Discussion

Our survey results indicate that traditional media still win the news battle. Television news for children in Flanders (*Karrewiet*) is watched by half of the respondents at least several times a week. Surprisingly, television news for adults is even more popular among children. Nearly two thirds of them watch it at least several times a week. Half of the respondents listen to radio news, from at least several times a week to almost every day. Far less attention is paid to the Internet. Only one third of the respondents answered affirmatively when asked whether they would surf the Internet (in general) when they were looking for something concerning the news. Nevertheless, nearly half of the respondents claim they would definitely visit a news site that would be specially developed for the young. In other words, if newsmakers would put greater effort into the development of a customised children's news site, more children would turn to online news in the future. Our results also indicate that despite the efforts of the newspaper industry and policy makers, such as the *Newspaper in the classroom* project (*Kranten in de klas*), printed newspapers are hardly read by children.

In our qualitative research we conducted fifteen in-depth interviews with Flemish children aged ten and eleven, eight boys and seven girls. During the interview, we used a semi-structured questionnaire to probe the news consumption of our participants, their motives for consuming news and their preference of topic. We also let them rank news media (TV, radio, the Internet and newspapers) while listing the pros and cons. Our results show that although the popularity of the Internet has risen throughout the years, children do not associate the Internet with news. When we asked them to rank four media according to their appropriateness as a news medium, most of the time, the Internet ended up last. The children in our quantitative and qualitative sample indicated they spent a lot of time surfing the net but did it mainly to keep in touch with friends, to play online games and to find information needed for school tasks.

Children show their interest in news especially by using traditional media. The television news is the leading medium here primarily because of the supremacy of images. These images clarify the content and children indicate that in this way, they had to put in less effort to understand what was going on in comparison to reading newspapers. Although television news was chosen as the most suitable news medium, participants also pointed to disadvantages. The newscast format has a certain duration that is often too long for children's concentration span. They get bored and the fixed sequence of the items irritates them. Newspapers benefit from the absence of this last aspect. The children put this medium second in the rankings because of its flexibility: you can read it wherever you want, exactly how long you want and you are in control of the sequence of the articles read. Radio ended third in the rankings and the Internet ended up last. It is striking that although children spend more and more time surfing the Web, the use of the Internet to gather news is not yet well established. None of our participants said he or she uses news sites often. Six children used them sometimes. One participant used it only once and six children said they knew about the existence of news sites or the possibility of reading newspapers online but that they never used it themselves. Two participants were not even aware of the existence of news sites.

The opportunities of online news are insufficiently known and used by children. Nevertheless the demands put forward by children concerning the news (flexible sequences, a lot of footage, a nearly unlimited supply, permanent updates, background information, different versions for different target groups, the use of uncomplicated language, etc.) are exactly the features of online news and the Internet scores a better grade on these aspects than the traditional news media. In conclusion we can say that the relationship between children and online news is currently suboptimal and needs new (governmental and non-governmental) initiatives for improvement in the near future. Initiatives for media literacy should develop the input aspect of online news consumption in the lives of children and realise their potential as young but nevertheless critical members of the public by involving them in producing output themselves. When reflecting on societal problems, reading and writing about them is crucial in the development toward active citizenship. Therefore, future research needs to analyse online news sites for children in a transnational and comparative way (for instance *Karrewiet* and *BBC Newsround*) and this comparison has to be extended to analyses of news sites for adults. Content analyses of this material can offer important insight into news coverage and the focus of children's news sites. The Flemish version of the children's news site *Karrewiet* offers relevant material with which to grasp children's attention, such as quizzes, files on elections, children's rights and road safety. Compared to the British site, however, the lack of a pre-moderated message board is a missed opportunity. An interactive tool to allow children to respond to the news they just read on the site or saw on the screen may induce discussion and reflection among peers and can increase understanding of news events.

Our study shows that the higher the educational level of the parents, the more often children surf the net for news or read newspapers. As the new round of digital divide studies has made clear, it is not so much about the haves and have-nots (Internet access) as it is about the new challenge to offer the same opportunities to develop searching skills to find the information one needs in an environment of overload of global data, and to think critically about what is presented and about the reliability of sources. As children of parents with a higher level of education tend to get more online news contact, this may lead to an increasing knowledge gap between children whose parents serve as *online news facilitators* and children who lack these initiators. Extra attention paid to online news and credibility of sources in the classroom may help to overcome this potential threat.

Innovative research, such as the use of "think aloud" protocol, in order to find out what children actually think while they are surfing a news site is another future challenge. Combined with screen and mouse-click registration this may produce useful information on elements that children find annoying, too difficult to grasp or parts of the news site they really do find interesting and want to know more about.

Next to the use children make of television news (both the adults and the children's versions), scales especially adapted to their age group need to be designed to measure the gratifications they are trying to satisfy by consuming news. Our findings show that adult news is watched a lot by children, but that the reasons behind this might need further exploration. Are they watching because it has become a habit, a sort of daily routine because their parents tune in, or are they forced to watch it in order to learn, and if so, is this a productive approach? Do children show intrinsic motivation in following the news to know what is going on or does it only entertain them in a certain way? These motives for news usage among children are even more important to focus on with respect to online news. Do children only happen onto news sites by accident or do they deliberately search for them? And does this aspect rapidly change over time in a way that might affect the future news consumption pattern of (young) people?

References

Ahlers, D. (2006). News Consumption and the New Electronic Media. *The Harvard International Journal of Press/Politics*, *11*(1), 1129-1152.

Baym, G. (2007). Emerging Models of Journalistic Authority in MTV's Coverage of the 2004 US Presidential Election. *Journalism Studies*, *8*(3), 382-396.

Betzel, L., Lauf, E., Negnborn, R., & Vosselman Bosh, J. (2009). *Concentratie en pluriformiteit van de Nederlandse media 2007*. Hilversum: Commissariaat voor de Media.

Broos, A. (2006). *De digitale kloof in de computergeneratie: ICT-exclusie bij adolescenten.* Leuven: K.U. Leuven School voor massacommunicatie research.

Buijzen, M., Van der Molen, J.H.W., & Sondij P. (2007). Parental Mediation of Children's Emotional Responses to a Violent News Event. *Communication Research*, *34*(2), 212-230.

Cantor, J. & Nathanson, A.I. (1996). Children's Fright Reactions to Television News. *Journal of Communication*, *46*(4), 139-152.

Costera Meijer, I. (2006). *De toekomst van het nieuws: hoe kunnen journalisten en programma-makers tegemoetkomen aan de wensen en verlangens van tieners en twintigers op het gebied van onafhankelijke en pluriforme informatievoorziening?* Amsterdam: Cramwinckel.

De Wit, J., Van der Veer, W., & Slot, M. (2001). *Psychologie van de adolescentie*. Baarn: HB uitgevers.

Donoso, V. (2007). *Adolescents and the Internet: Implications for Home, School and Social Life*. Leuven: K.U. Leuven School voor massacommunicatie research.

Drew, D.G. & Reeves, B.B. (1980). Children and Television News. *Journalism Quarterly*, *57*(1), 45-114.

Evard, M. (1996). Children's Interests in News: Online Opportunities. *IBM Systems Journal*, *53*(3.4), 417-430.

Gross, E.F. (2004). Adolescent Internet Use: What We Expect, What They Report. *Applied Developmental Psychology*, *25*, 633-649.

Heene, J., Vossen, I., Van Damme, D., & Buysse, P. (1999). *Cultuurinitiatie door massamedia, Experimentele ontwikkeling van een vormingspakket voor het einde van het basisonderwijs*. Gent: Vakgroep Onderwijskunde.

Lenhart, A., Madden, M., Rankin Macgill, A., & Smith, A. (2007). *Teens and Social Media (Pew Internet and American Life Project Report)*. Retrieved January 10, 2008, from: www.pewInternet.org/pdfs/PIP_Teens_Social_ Media_Final.pdf.

Ngyen, A. (2008). *The Penetration of Online News: Past, Present and Future.* Saarbrucken: VDM Verlag.

Patterson, T.E. (2008). Young People Flee from the News, Whatever the Source. *Television Quarterly*, *38*(2), 32-35.

Pew Research Center (2008). *Audience Segments in a Changing News Environment. Key News Audiences Now Blend Online and Traditional Sources. Pew Research Center Biennal News Consumption Survey*. Retrieved June 5, 2010, from http://people-press.org/http://people-press.org/files/legacy-pdf/444.pdf.

Raeymaeckers, K. (2002). Young People and Patterns of Time Consumption in Relation to Print Media. *European Journal of Communication*, *17*(3), 369-383.

Raeymaeckers, K. (2004). Newspaper Editors in Search of Young Readers: Content and Layout Strategies to Win New Readers. *Journalism Studies*, *5*(2), 221-232.

Roe, K. (2006a). The Digital Divide in the Twenty-first Century: An Introduction Source. *Poetics*, *34* (4.5), 219-220.

Roe, K. (2006b). Television, Viewer Age. In J.J. Arnett (ed.), *Encyclopaedia of Children, Adolescents and the Media* (pp. 820-822). Thousand Oaks Ca: Sage.

Rogers, E. (1995). *Diffusion of Innovations*. New York: Free Press.

Smith, S.L. & Wilson, B.J. (2002). Children's Comprehension of and Fear Reactions to Television News. *Media Psychology*, *4*(1), 1-26.

Sparks, G. (2006). *Media Effects Research: A Basic Overview.* London: Thomson.

Valkenburg, P. (2008). *Beeldschermkinderen. Theorieën over kind en media.* Amsterdam: Boom.

Vandebosch, H., Dhoest, A., & Vandenbulck, H. (2009). *De inhoud, vorm en programmering van jongerenjournaals: lessen uit de evaluatie van Zoom (VTM) en Jam (VT4)*. Nijmegen: Etmaal van de communicatiewetenschap.

Van den Bulck, J. & Van den Bergh, B. (2000). The Influence of Perceived Parental Guidance Patterns on Children's Media Use: Gender Differences and Media Displacement. *Journal of Broadcasting & Electronic Media, 44*(3), 329-348.

Van den Bulck, J. (2006). Television News Avoidance: Exploratory Results from a One-year Follow-up Study. *Journal of Broadcasting & Electronic Media, 50*(2), 231-252.

Walma Van der Molen, J.H., Valkenburg, P.M., & Peeters, A.L. (2002). Television News and Fear: A Child Survey. *Communications, 27*(3), 303-317.

Positive Parents Taking Action

Parental Mediation of Children's Digital Game-playing

Willem OOSTING, Yvonne A.W. DE KORT
& Wijnand A. IJSSELSTEIJN[1]

Introduction

Ever since the 1970s when digital games were introduced, the digital game market has been growing (ISFE, 2009). In fact, digital game-play today is so popular that it has become a mainstream activity for contemporary youth and adolescents (ESA, 2009; Nielsen Games, 2008; Pratchett, 2005; Rideout, Foehr, & Roberts, 2010). Games, in general terms, are hedonic activities engaged in by players voluntarily for recreational purposes. Salen and Zimmerman (2003) further specified them as being systems in which players engage in goal directed behaviour (e.g. obtaining a certain object, winning an artificial conflict), defined by rules, that results in a quantifiable outcome. Digital games are now ubiquitous, and offered on a range of devices including PCs, consoles, handheld devices, smart-phones and PDAs. British statistics indicate that below the age of eleven, the penetration of games has now reached 100 per cent (in the age group just above, 11-15-year-olds, it is 97%), implying that every child plays digital games at least occasionally (Pratchett, 2005). The current chapter regards an empirical investigation of parental mediation in relation to their children's consumption of such digital games.

Parental Mediation

Market research shows that the majority of modern parents take an interest in their children's game consumption patterns. In addition, game

1 This work was supported by the Games@Large project under the Information Society Technologies priority of the European Union's Sixth Framework Programme [grant number IST-1-038453-IP]. Additionally we would like to thank the gamers and both experts: Huub De Hesselle (Msc. PDEng.) and Antal Haans (PhD) for their help in developing the behaviour based attitude scale. We furthermore want to express gratitude towards the participating parents and families providing us with their time and effort to fill out the questionnaire, forming the foundation of this research.

advisory systems such as PEGI[2] (EU) and ESRB[3] (North America) are valued tools used by parents to base game choices on (Nikken, Jansz, & Schouwstra, 2007). However, recent qualitative research shows that parental mediation of digital game behaviour is a complex phenomenon and varies between parents (Kutner, Olson, Warner, & Hertzog, 2008; Oosting, IJsselsteijn, & de Kort, 2008). It encompasses more than simply allowing children to either buy or play certain games on the basis of an age rating. In fact, parents use various methods to direct their children's digital game behaviour according to their norms and values (Oosting *et al.*, 2008). Previous quantitative research (Nikken, 2003) provided evidence that mediation strategies for gaming resemble commonly applied strategies in parental mediation of television viewing. Traditionally these TV mediation strategies are grouped into three categories (Nathanson, 2002; Valkenburg, Krcmar, Peeters, & Marseille, 1999). In the domain of gaming, similar categories emerged and were labelled in correspondence to television mediation; (i) *restrictive mediation* – setting explicit limits on for example time or certain game types, (ii) *active mediation* – discussing or explaining aspects of games, and (iii) *co-play* – playing games together (Nikken & Jansz, 2006). Research on mediation of computer use in the classroom, and TV viewing at home, has demonstrated the practical importance of making the right mediation choices, as they can yield both beneficial (e.g. Klein, Nir-Gal, & Darom, 2000) and detrimental (e.g. Nathanson, 2002) effects in certain situations. However, at present, empirical insight is lacking into the underlying psychological mechanisms that drive parents' choices in *digital game* mediation.

General Parenting Styles

The choices parents make in mediating their children's game behaviour take place in a larger normative context of their general dispositions towards raising children. During the 1970s Baumrind (1971) distinguished three general parenting styles, largely based on the parents' vision on authority. These often cited categories are: (i) *permissive parenting* – stimulating children to make their own decisions, (ii) *authoritarian parenting* – demanding immediate obedience and (iii) *authoritative parenting* upholding directive regulations but at the same time

---

2 A twofold rating system adapted from the Dutch *Kijkwijzer* television content rating system; on the one hand using pictorial labels to warn parents for possible harmful content and on the other hand an age-rating meaning no harm is to be expected for children from that age onwards.

3 Entertainment Software Rating Board classification system similar to PEGI, but with a more elaborate set of rating categories and game content *descriptions* instead of *icons*.

discussing them while taking children's wishes into account. During that same period, the Family Communication Patterns (FCP) model was developed in order to describe how families reach a shared social reality, which – according to the authors – is needed for a family to smoothly function (Koerner & Fitzpatrick, 2004; McLeod & Chaffee, 1972). The model describes two underlying dimensions, one of which is called the *socio orientation* and is somewhat similar to Baumrind's authority scale. The second dimension, termed *concept orientation* focuses on the type of family communication. Koerner and Fitzpatrick (2004) later redefined these two FCP dimensions into respectively: (i) *conformity orientation* – the extent to which a parent strives towards shared consensus on norms and values, and (ii) *conversation orientation* – the extent to which family members are encouraged to take part in discussions about various topics.

Research in the domain of television mediation has demonstrated that general parenting styles may have an impact on the specific routes parents take to influence their child's viewing behaviour (e.g. Fujioka & Austin, 2002; Warren, Gerke, & Kelly 2002). Fujioka and Austin (2002) studied how FCP parenting styles relate to active mediation from a communication perspective. Their telephone survey showed that parental concept (communication) orientation is tied in with all forms of active mediation, that they defined as: positive mediation (i.e. reinforcement), negative mediation (i.e. negative reinforcement) and critical discussion. The socio (conformity) orientation only correlated with positive active mediation. In addition, they linked this orientation to both co-viewing and the enforcement of TV rating restrictions.

As some of the parental mediation strategies in relation to gaming appear to be similar to those in relation to TV viewing (e.g. Nikken, 2003; Nikken & Jansz, 2006), general parenting styles may have a similar influence on parental mediation in the domain of digital gaming. In particular, restrictive strategies are similar in nature. Therefore, in line with the findings of TV research, we hypothesise that authoritarian parents (in Baumrind's terms), or those with a high conformity orientation (in Koerner and Fitzpatrick's terms), employ more restrictive strategies than others. In addition, authoritative parents – placing emphasis on both family conformity and communication – are expected to engage in more parent-child discussion about the games (i.e. active mediation), whereas permissive parents – with a low conformity orientation – may have a more laissez-faire attitude, mediating less than others altogether.

Digital Game Attitudes

In addition to general parenting styles, parental dispositions towards the medium itself may also influence mediation choices. TV related research (e.g. Warren, 2003; Warren *et al.*, 2002) provides evidence for a relationship between parents' attitudes towards TV and the mediation choices they make when their children watch TV. More specifically, their research shows that parents' attitudes have a negative correlation with both restrictive and instructive (active) mediation. Such relations might also hold for digital game mediation. It seems intuitively plausible that parents who generally hold strong positive attitudes towards games might use different mediation strategies than parents that hold less positive ones. In line with the findings of Warren *et al.* (2002), we expect that parents with a strong positive view of digital games will use less restrictive mediation than parents whose attitude is not as positive. Warren *et al.* found no significant correlation between attitudes and co-viewing. However, qualitative research specifically addressing parental mediation of digital game behaviour provided qualitative evidence that co-playing and co-viewing might fundamentally differ from each other as the latter allows for a much more passive *lean back* mode of interaction, whereas co-playing requires an active *lean forward* stance and an investment of time in activities most parents might not have selected themselves (Oosting *et al.*, 2008). So, despite the fact that Warren did not find a relation between parents' attitudes and co-viewing in the field of television mediation, we might still find that parents with strong positive attitudes with respect to gaming play more digital games with their children than those having less positive attitudes.

Perceptions About the Possible Effects of Digital Gaming on Children

Public concerns around the issue of media effects have a long history, dating back well before digital technology. Such concerns mostly focus on moral issues, most obviously sex and violence. Indeed, the debate on digital games is no exception. Both popular media and political discourse (e.g. Brekke & Lepre, 2007; Dobson, 2006) have frequently referred to digital games as a controversial leisure activity, from which young children need to be protected. Unfortunately, parents cannot take guidance from a systematic and balanced scientific treatment of this topic. In academic literature, the debate on whether and to what extent games have either positive or negative effects has been polarised and is largely unresolved (e.g. Anderson & Bushman, 2001; Charlton & Danforth, 2007; Ferguson, 2007), partly due to methodological disputes and partly due to differences in theoretical perspectives. Qualitative research, however, indicates that parental perceptions about the effects of digital gaming do not necessarily focus on moral concerns, but have a much broader scope, including concerns on reduced social interactions within the family and reduced educational achievement, yet also including positive views on the variety of skills developed while playing digital games, and their importance as a source of fun and leisure (Kutner *et al.*, 2008).

It is important to note that the parental attitudes discussed in the previous section refer to general dispositions towards digital games as part of the parents' *own* media consumption patterns. These attitudes do not necessarily coincide with the specific views parents hold regarding the effects that games can have on their children. As an example, parents may feel that playing a game that contains sexually explicit content is an acceptable and enjoyable pastime for *them*, but would not want to expose *their young children* to the same game for fear of potential negative psychological effects. Thus, one's general attitude towards a game may be quite different from one's perception of its potential effects on children. One may expect that parents who hold strong views on the potential negative effects (or perceived risks) of games are more likely to adopt a highly restrictive mediation style. However, very little empirical data are available on how such perceptions may affect parental mediation choices.

Current Study

In the current study, we first investigate and quantify the possible relations between parenting styles and parents' attitudes towards digital games on the one hand, and their mediation strategies regarding their children's game activities on the other. Based on the similarities in mediation strategies regarding TV viewing and digital gaming, we would expect parenting styles to have similar effects on digital game mediation, as found for TV viewing mediation. Then, in addition to these predictors of parental mediation, this study also takes into account parents' likelihood perceptions regarding possible gaming consequences for their children. We investigate how probable parents think various possible gaming effects are and subsequently relate such perceptions to parental mediation by adding these to the predictive model. Our study's overarching goal is to aim for inroads towards understanding the psychological mechanisms underlying parental mediation behaviour that may be utilised to support the future development of appropriate tools to aid parents in making informed mediation choices.

Method

The questions to be answered are all quantitative empirical questions. As such, in order to be able to make significant inferences, a large body of data is needed. When the Internet usage penetration rate is high, as it is in the Netherlands (Miniwatts Marketing Group, 2009), online surveys provide an efficient and reasonably representative means to gather such large amounts of data. This method has been deployed previously in the area of digital game mediation, and with good results (e.g. Nikken & Jansz, 2006). Hence, a web survey was designed (Schmitz, 2009), placed on-line for several months and participants were asked to fill it out in the convenience of their own home. However, they were instructed not to consult others while doing so and thus fill out the questionnaire by themselves. Completing the questionnaire took about 35 minutes on average.

Participants

Parents of children between three and twenty-three years, still living at home, were invited via a participant database, contacted via a snowball sampling technique, or recruited directly during public family events at the university. In addition, flyers were handed out in shopping malls and village/city centres directing parents to the web address hosting the survey. During the recruitment phase it was stressed that their disposition towards digital games was no selection criterion (hence: also parents that did not allow their children to play digital games were invited to participate). As an external incentive to participate, all parents were offered free participation in a raffle of theme park tickets.

In total, 142 Dutch parents (average age = 42, $SD_{age} = 6.3$, 60% female) of children (average age = 10, $SD_{age} = 4.8$, 35% female) filled out the web survey. About half of the participating parents played digital games themselves on a weekly basis or more; on average spending 2.3 hours a week ($SD_{game\text{-}time} = 6.7$). In filling out the questionnaire they were asked to take in mind the child with his or her birthday coming up the soonest (to prevent a possible selection bias). Almost half of them reported this child to play digital games each day; the other half reported their children to play weekly or less.

Measures

The survey consisted of four main sections. These were designed to measure: i) mediation strategies, ii) parenting styles, iii) game attitude, and iv) game effect likelihood perception. Additional parameters were also measured, but not reported in the current manuscript as they fall outside the scope of this chapter.

Mediation Strategy

To measure the extent to which parents use the three different types of mediation strategies, a questionnaire developed by Nikken and Jansz (2006) was used. The questionnaire allows participants to indicate how often they employ 15 mediation acts on a four-point scale, ranging from "Never" to "Often". Because a principal components analysis (PCA) investigating the two-, three- and four-factor solutions did not yield interpretable dimensions and no other tools based on sound theoretical principles were available, the original scales as determined by Nikken and Jansz (2006) were employed as a best practice. Hence, participants' parental mediation strategy scores for each scale were calculated by averaging the non-weighted item scores for all items within each scale. This rendered subscales with acceptable to good internal consistencies: co-play (3 items, $\alpha = .88$), active mediation (5 items, $\alpha = .78$) and restrictive mediation (5 items, $\alpha = .79$).

Parenting Styles

The revised Parental Authority Questionnaire (PAQ-r) was used to measure parenting styles (Reitman, Rhode, Hupp, & Altobello, 2002). Parents were asked to indicate to what extent they agreed with 30 general statements about raising children on five-point Likert scales, ranging from "Strongly Disagree" to "Strongly Agree". Again a PCA was performed; this time searching for *three* scales representing parenting styles, with the same rotation (Oblimin) and items having communalities less than .3 and/or cross-loadings higher than .3 being discarded. Unfortunately, the three-factor structure yielded factors that were difficult to interpret in terms of Baumrind's theory. We therefore discarded this factor structure and explored a two-factor structure, which corresponded closely to the two dimensions defined by Koerner and Fitzpatrick (2004). Participants' mean scores were subsequently calculated by averaging the non-weighted item scores per dimension: conformity- (4 items, $\alpha = .68$) and communication orientation (3 items, $\alpha = .54$). These dimensions showed a modest negative correlation ($r = -.25$).

Game Attitude

No scale was available that would allow us to measure parents' attitude towards digital games in general (i.e. reflecting also their own use of games, and not just their children's). Therefore a new scale was developed to measure digital game attitudes. We took the approach of including behaviour-based items, as reporting on actual behaviours is generally more reliable than asking people's opinions. We employed Campbell's approach in this attitude scale development (e.g. Campbell, 1963; Kaiser, Byrka, & Hartig, 2010). According to this paradigm, there is a direct relationship between a person's attitude or motivation to attain a particular goal and the behavioural means he or she employs to attain that goal. A list of 34 items was generated from a brainstorm session with gamers about behaviours that they think are caused by a latent game attitude. The game-related behaviours varied in likelihood or "difficulty". For instance: "How often do you play a (free) digital game?" and "how often do you keep things from others in order to be able to play digital games longer". The probability of any person engaging in a specific activity on this list depends on (a) a person's motivation to play games, and (b) the costs involved in performing the particular activity. This rationale can be mathematically formalised with the Rasch model (for mathematical details see: e.g. Bond & Fox, 2007). The set of items was developed to create a suitable, workable and sensitive measurement tool. Ten items followed a dichotomous yes/no format; another set of twenty-four items had five-point frequency response scales ranging from never to very often. The latter were dichotomised during scale calibration.

Scale calibration resulted in an item difficulty distribution that had an optimal overlap with the distribution of parents during the Rasch model test. Analysis was performed using the Facets software iteratively (Linacre, 2006). This package employs a joint maximum likelihood procedure for estimating parents' motivation to play digital games and to estimate the difficulty of each of the game related behaviours. The units of these estimates are called logits. The scale had a person separation reliability of .90. This measure indicates that there is good person separation (there are persons scoring high as well as persons scoring low), and that the order of persons in terms of their motivation is consistent (Bond & Fox, 2007). It matches with a classical definition of reliability (e.g. Cronbach's alpha). More than 92% of participants fitted the model well. The overall fit for the 34 items was found to be sensible as well: M(MS) = 0.98, SD(MS) = 0.2.[4]

Perceived Risks and Benefits of Gaming

Parents were presented 30 effects potentially resulting from playing digital games. These were adapted from scientific literature (e.g. Nikken & Jansz, 2006), debates in popular media, but mainly distilled empirically from focus groups, contextual inquiries and diary studies that were run in an earlier study (Oosting, IJsselsteijn, & de Kort, in preparation). Examples of listed possible effects are: "development of spatial aptitude" and "development of obesity". For each item, presented in a random order, parents indicated, on a 5-point scale how likely they perceived the various effects regarding their own child (ranging from very unlikely to very likely).

A Principal Component Analysis (PCA) with a forced two-factor, oblimin rotated solution resulted in two, only slightly correlated (r =.20) scales. One measures parents' perceived likelihood of gaming benefits (15 items; α =.90), the other of gaming risks (15 items; α =.87).

4 MS-values reflect the relative discrepancy in variation between model prediction and observed data. Such values are ideally 1.00. A MS-value of 1.20, for example, indicates 20% more variation in the data than what is predicted by the Rasch model.

Results

Determinants of Parents' Mediation Choices

Two parenting style variables were computed, representing parents' conformity and conversation orientation respectively. Participants scored around the neutral point on conformity orientation (M=3.3, SD=0.65) and slightly higher on communication orientation (M=3.9, SD=0.50). See also Table 1. Furthermore, each parent's gaming attitude was calculated (M=-0.69, SD=1.92).

Regarding the application of mediation strategies, paired sample t-tests showed that parents significantly use more restrictive mediation (M=2.6, SD=0.76) than active mediation (M=2.4, SD=0.71; t=4.0, p<.001, df=141) or co-play (M=2,1, SD=0.85; t=6,5, p<.001, df=141). Active mediation is in turn used significantly more frequently than co-play (t=3.6, p<.001, df=141).

Table 1: Respondents' scores on parenting style, mediation strategies and gaming attitude (n=142)

	Minimum	Maximum	Mean	Std. Deviation
Conformity Orientation	1.00	4.50	3.30	.65
Communication Orientation	2.33	5.00	3.86	.49
Restrictive Mediation	1.00	4.00	2.54	.76
Active Mediation	1.00	3.80	2.36	.71
Co-Play	1.00	4.00	2.11	.84
Game Attitude	-5.78	3.47	-.68	1.91

Bivariate correlations were subsequently calculated in order to investigate possible significant relationships between general parenting styles, game attitude and mediation strategies (See Table 2).

Table 2: Bivariate correlations between predictors and mediation strategy scores (n=142)

	Conformity Orientation	Comm. Orientation	Restrictive Mediation	Active Mediation	Co-Play	Game Attitude
Conformity Orientation	1	-.25**	.23**	.05	.22**	.05
Communication Orientation	-.25**	1	.12	.19*	-.03	.09
Restrictive Mediation	.23**	.12	1	.72**	.52**	.24**
Active Mediation	.05	.19*	.72**	1	.45**	.26**
Co-Play	.22**	-.03	.52**	.45**	1	.42**
Game Attitude	.05	.09	.24**	.26**	.42**	1
Perceived Risks	.13	-.05	.07	.12	-.10	.04
Perceived Benefits	.05	.21*	.19*	.28*	.25**	.31**

**. Correlation is significant at the .01 level (2-tailed); *. Correlation is significant at the .05 level (2-tailed).

Linear regression analyses were performed to predict the three mediation strategies with the two dimensions of general parenting style and game attitude. Results showed that for restrictive mediation, all three predictors were significant: restrictive mediation was related to a stronger conformity orientation (Beta=.26, $p<.001$), a stronger communication orientation (Beta=.17, $p<.05$), and a more positive game attitude (Beta=.22, $p<.02$). The predicted variance was modest (R^2=13%). Two significant predictors emerged for co-play: more co-play was related to a stronger conformity orientation (Beta=.21, $p<.02$), and a more positive game attitude (Beta=.42, $p<.001$; R^2=22%). Active mediation was predicted by a stronger communication orientation (Beta=.19, $p<.03$) and a more positive game attitude (r=.24, $p<.01$; R^2=10%). A corresponding model was constructed to depict how parenting styles and parents' game attitude relate to mediation strategies (see Figure 1).

Figure 1: Partial correlations between parenting style dimensions: conformity orientation and communication orientation, game attitude and mediation strategies (crossing correlations being partial)

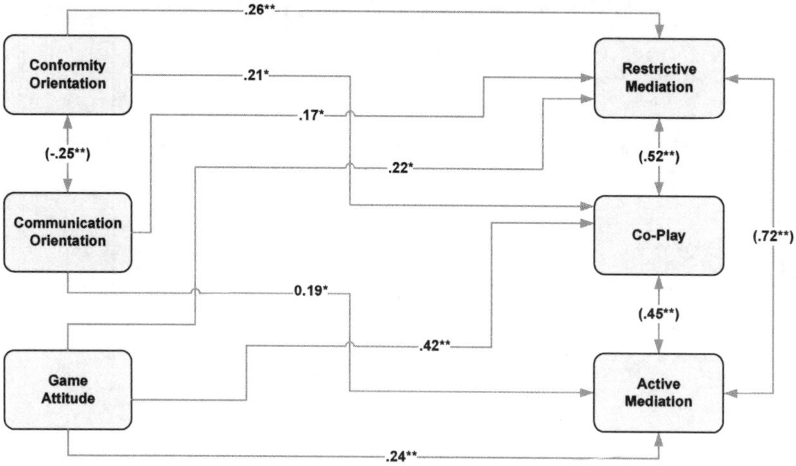

**Correlation is significant at the .01 level and * at the .05 level

Extending the Model with Parents' Perceptions of Possible Gaming Effects

In the previous paragraph we predicted mediation strategies with general parenting styles and parents' own attitude towards digital games, similar to what had been explored for TV viewing. In the present paragraph we explore the additional predictive validity of a model that includes parents' perceptions of the risks and benefits of digital games for their children. On average, participants' scores regarding the *risks* of digital gaming leaned towards "unlikely" (M=2.11, SD=0.57). Paired sample T-test (repeated measure) analyses demonstrated that participants significantly (t=19.4, p<.001, df=141) scored digital gaming *benefits* to be more likely, on average scoring these higher than the neutral category (M=3.31, SD=0.64). The bivariate correlations of these two variables with the dependent variable and remaining predictors are reported in Table 3.

Table 3: Perceived likelihood scale scores for risks and benefits (n=142)

	Minimum	Maximum	Mean	Std. Deviation
Perceived risks	1.00	4.07	2.11	.57
Perceived benefits	1.00	5.00	3.31	.64

Figure 2: Partial correlations between parenting style dimensions: conformity orientation & communication orientation, game attitude, perceived risks & benefits and mediation strategies (crossing correlations being partial)

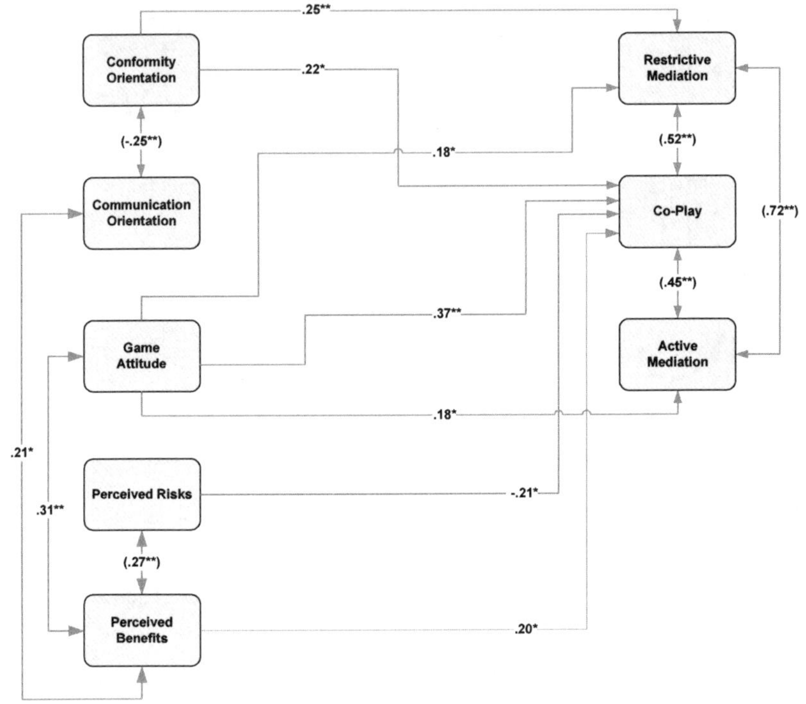

**Correlation is significant at the .01 level and * at the .05 level

The initial model was extended with the perceived risks and benefits (see Figure 2). Multiple regression analyses were again performed for each of the three mediation strategies. Results showed that for restrictive mediation, two predictors remained significant: restrictive mediation was related to a stronger conformity orientation (Beta=.25, p<.001) and a more positive game attitude (Beta=.18, p<.04). The contribution of communication orientation bordered on significance (Beta=.16, p<.07). The predicted variance improved only slightly (R^2=14%). In this new model four significant predictors emerged for co-play: as in the first model, more co-play was related to a stronger conformity orientation (Beta=.22, p<.02), and a more positive game attitude (Beta=.37, p<.001). In addition, both perceived risks and benefits predicted the occurrence of co-play: a higher likelihood of risks was related to lower frequencies of co-play (Beta=-.21, p<.02), whereas a higher likelihood of benefits was related to more frequent co-play (Beta=.20, p<.03). The predicted variance for co-play improved (R^2= 27%). Active mediation was predicted by a more positive game attitude (r=.18, p<.03); the contribution of perceived benefits bordered on significance (Beta=.17, p=.053). The predicted variance again improved modestly (R^2= 14%).

Discussion

The web survey research described in the current chapter first and foremost set out to explore how possible predictors relate to parental mediation choices regarding their children's digital game behaviour. We explored whether the relationships found in television mediation research also occur in the domain of digital gaming. In line with these studies, *general parenting styles*, and *parents' attitudes* towards digital game-play were used to predict three types of mediation strategies employed by parents: restrictive mediation, active mediation, and co-play. Parents' conformity- and communication orientation (parenting style) were measured by interpreting PAQ-r item scores (Reitman *et al.*, 2002) from a communication research perspective (McLeod & Chaffee, 1972; Ritchie & Fitzpatrick, 1990). In addition, a game attitude scale was constructed according to Campbell's paradigm by fitting behaviour-based items to a Rasch model (Bond & Fox, 2007). The scale employed various behaviours related to their game motivation, ranked by item difficulty to predict parents' latent game attitude. Parents' scores on these scales were related to parental mediation strategies, measured by a questionnaire designed to measure appliance of restrictive mediation, active mediation and co-play (adopted from Nikken, 2003). Using multiple linear regression analyses unique relationships were calculated.

In a second iteration the model was extended by taking into account parents' perceived likelihood of possible positive and negative effects of digital games on their children. To this end two additional scales were formed, one measuring the subjective likelihood of digital gaming benefits and another measuring the subjective likelihood of risks.

In considering the data, what catches the eye first is the fact that the three parental mediation strategies showed positive and rather strong inter-correlations. This implies that parents in general do not clearly prefer and select one over the other strategy, but rather employ various strategies in dealing with their children's gaming behaviour (or conversely, employ no strategy at all). As an example, although perhaps counterintuitive: the parents that engaged in restrictive mediation – i.e. restricted game time or certain games – were more likely to also engage in co-play. These findings also reflect the rather nuanced views parents have of digital games and how they affect their children's development. This is further corroborated by our findings on parents' perceived likelihood of beneficial and detrimental consequences of games, which will be discussed later.

The first set of regression analyses indicated that the first parental mediation strategy – restrictive mediation – was related with all three predictors: the two dimensions of parenting style and parents' attitude towards games. Moreover, each of the three predictors showed a positive correlation, indicating that parents with a stronger conformity orientation, a stronger conversation orientation, and a more positive attitude towards games restricted their children's gaming behaviour more frequently. Especially this latter finding was unexpected. We will return to this topic at the end of this section. Co-play was initially predicted by a stronger conformity orientation of the parents and again, a more positive game attitude. These findings were in line with our hypotheses. Lastly, active mediation showed significant correlations with a stronger conversation orientation and more positive game attitude. Again, these findings were in line with our hypotheses.

Before discussing the role of our predictors in more detail, we will first consider the second set of regression analyses. In addition to the general parenting styles and the parents' general attitude towards games, we now entered indicators of the risks and benefits parents may perceive digital games carry with them. In previous television research parents' attitudes were often operationalised with items measuring parental beliefs about the potential harmful effects of media exposure on their children (e.g. Warren, 2003; Warren *et al.*, 2002). In the current study we measured parents' general attitude towards the medium itself. In addition we assessed parental beliefs about opportunities and risks of digital games for their children. Because parental mediation is generally used to the benefit of the children (Ritchie & Fitzpatrick, 1990) we hypothesised that such parental beliefs about digital gaming effects on their children would play an important role. Moreover, given the fact that these perceptions related to the exact topic at hand – their children playing digital games – or in other words that these predictors were more close to the variables we were trying to predict (i.e. the strategies employed to mediate these behaviours), we expected these variables to show up as strong predictors of mediation strategies – stronger even than the predictors already in the model, as these were of a more distant type (parenting in general, digital games in general). This turned out not to be the case: the new predictors showed significant relations with co-play (perceived likelihood of benefits contributed to co-play, whereas higher likelihood of risks decreased the frequency of co-play), but not with the other mediation strategies (although the relation between subjective benefits and active mediation bordered on significance). Simultaneously, the relationships between the communication orientation and mediation strategies disappeared. The total predicted variance showed modest increases when these perceptions were included in the model.

The initial and extended model showed that parents' orientation towards family conformity has a moderately positive correlation with both parental restrictive behaviour and co-play. This is in line with findings in TV research domain where Fujioka and Austin (2002) for instance demonstrated similar results. The positive correlation between parents' orientation towards family communication and active mediation found in the initial model was also in line with TV research. Yet, in addition, a positive relationship between parents' attitudes towards digital game behaviour and all three parental mediation strategies emerged in our research. This is in stark contrast to TV mediation research where an inverse relationship was found between parents' TV attitudes and both restrictive and instructive (active) mediation (Warren *et al.*, 2002). They reported no relation between parents' attitudes and co-viewing. Overall, although indeed parenting styles and perceived risks and benefits of games modestly predicted parents' use of mediation strategies, the strongest predictor in the current model, for each of the mediation strategies, was parents' general game attitude.

How can we explain this finding? Why do parents with a stronger and more positive attitude towards games – *for themselves, not their children* – engage in more mediation, be it co-play, active, or restrictive? We briefly considered that perhaps parents who had played digital games, or were still gaming, were more familiar with the medium and hence more aware of the risks (and benefits) of such games and therefore employed more mediation strategies to protect and guide their children. However, bivariate correlations show that general parental attitudes are not related to the perceived likelihood of risks of games. The correlation with perceived benefits was very modest. Parents who enjoy playing games themselves see just as many risks for their children in gaming as parents who do not like digital games. We must also assume that their children's welfare and development is equally important to both groups of parents. Then what is the relevant difference between these groups? Although this is impossible to prove with the current dataset, a potential explanation of the differences may be parents' media literacy in the area of digital gaming. Digital gaming is a relatively young media phenomenon, and many parents state they have little knowledge about it (Oosting *et al.*, 2008), whereas parental TV knowledge is likely to be both more extensive and more uniformly distributed across the population. A positive attitude on a behaviour-based scale signals that these parents are more likely to be more involved with gaming in general and to put in more effort in digital games than other parents. Due to these experiences, they may be more literate with respect to games and therefore better equipped to formulate restrictions in terms of playing time or suitable games and genres, to play games (and enjoy it) with their children, and to discuss the ins and out, pros and cons of games and playing games. Whether this indeed is the explanation behind our findings should be explored in future research.

A few words of caution in interpreting the current data: only Dutch parents participated in the current study, whereas parents from other countries or cultures might hold different views. Also, since the penetration rate of Internet usage is very high in the Netherlands, we felt safe in assuming an online survey would not bias the dataset in any meaningful way. The fact remains that in employing this methodology, researchers do not have full control over who fills out the questionnaires, and whether these participants perfectly match the sample criteria. Moreover, the findings on parenting styles are based on measurements from an itemset that was not specifically developed to be interpreted from a communication research perspective having moderate internal consistency. For future research in this area, it is advisable to consider including additional questionnaires that specifically tap into the dimensions of conformity and communication orientation. An example of such a questionnaire is the revised Family Communication Pattern (FCP) questionnaire (Ritchie & Fitzpatrick, 1990). Lastly, the children of the parents in our research sample spanned an age range of almost 20 years. As parenting styles may well vary with the age of the children, we would have liked to explore such trends in the current dataset. Unfortunately, the size of our sample would not allow such explorations. In fact, in view of this potential diversity in parenting styles across children's age, the sample size of the current study is already quite modest.

Conclusion

This study presented a first quantitative exploration of factors that are thought to underlie parental mediation of digital games. Based on earlier work reported in TV mediation literature, we hypothesised that parenting styles and parental attitude towards games (as part of parents' own media consumption pattern) would be good predictors of parental mediation with regard to their children's gaming behaviour. Moreover, we also expected effects of perceived risks and benefits of games on parents' choice of mediation strategies. Our findings were novel and surprising, only partially supporting our initial expectations. Perceived risks and benefits predicted parental mediation behaviour, but only modestly so, and only in relation to co-play. In addition, parents with a higher tendency to impose norms and values onto their children set more limits but also played more digital games with their children (i.e. co-play). The strongest predictor of mediation behaviour, however, was parents' general attitude towards games: parents who had a more positive attitude towards games also tended to engage in more frequent mediation of their children's game behaviour, employing the full range of mediation strategies in the process. This effect was not related to an increase in perceived risks of digital games, but may instead point to a key role of media literacy in parental mediation behaviour. With a larger set of skills and knowledge regarding digital games at their disposal, it becomes easier for parents to mediate different gaming behaviours in different ways, carefully choosing and tailoring the right mediation strategy as required, at times sharing and discussing experiences of game-playing, at other times applying a more firm, restrictive approach

The nature of digital games is changing rapidly. Over recent decades, games have become complex, extensive, and ubiquitous game worlds, with high levels of visual realism, and online play modes. Games, especially online ones, are by no means self-contained, linearly evolving stories, like films, but rather interactive experiences that are, at least partly, self-created by the people playing the games. Parents clearly have a central role to play in guiding their children's game consumption patterns, although this is by no means an easy task, especially for someone not familiar with today's digital games. As suggested by our results, it appears to be the parent's own level of media literacy that plays a major role in choosing the appropriate mediation strategy. Thus, developing forms of media literacy that promote a critical understanding of games and the role they play in the lives of their children can empower parents to be more actively engaged with their children's media education. The development of a more systematic and informative classification and labelling of games should certainly be part of such an agenda, extending current parental mediation tools, such as filters or age verification systems, as well as content advisory systems such as PEGI and ESRB.

References

Anderson, C.A. & Bushman, B.J. (2001). Effects of Violent Video Games on Aggressive Behavior, Aggressive Cognition, Aggressive Affect, Physiological Arousal, and Prosocial Behavior: A Meta-Analytic Review of the Scientific Literature. *Psychological Science*, *12*(5), 353-359.

Baumrind, D. (1971). Current Patterns of Parental Authority. *Developmental Psychology Monograph*, *4*(1), 1-103.

Bond, T.G. & Fox, C.M. (2007). *Applying the Rasch Model: Fundamental Measurement in the Human Sciences* (2nd ed.). London: Lawrence Erlbaum Associates Inc. Publishers.

Brekke, A. & Lepre, C.R. (2007). *Burned by "Hot Coffee"? Parental Trust and Usage of the Video Game Rating System*. San Francisco: Annual Meeting of the International Communication Association.

Campbell, D.T. (1963). Social Attitudes and Other Acquired Behavioral Dispositions. In S. Koch (ed.), *Psychology: A Study of a Science* (pp. 94-172). New York: McGraw-Hill.

Charlton, J.P. & Danforth, I.D. (2007). Distinguishing Addiction and High Engagement in the Context of Online Game Playing. *Computers in Human Behavior*, *23*(3), 1531-1548.

Dobson, J. (2006). *EU Commissioner Questions Rule of Rose Content*. Retrieved 29 January, 2010, from http://www.gamasutra.com/php-bin/news_index.php?story=11738.

ESA. (2009). *Essential Facts About the Computer and Videogame Industry 2009: Sales, Demographic and Usage Data*. Retrieved 13 January, 2010, from http://www.theesa.com/facts/pdfs/ESA_EF_2009.pdf.

Ferguson, C. (2007). The Good, the Bad and the Ugly: A Meta-analytic Review of Positive and Negative Effects of Violent Video Games. *Psychiatric Quarterly*, *78*(4), 309-316. doi:10.1007/s11126-007-9056-9.

Fujioka, Y. & Austin, E.W. (2002). The Relationship of Family Communication Patterns to Parental Mediation Styles. *Communication Research*, 29(6), 642-665.

ISFE. (2009). The World Market by Segment. Retrieved March 19, 2010, from http://www.isfe-eu.org.

Kaiser, F.G., Byrka, K., & Hartig, T. (2010). Reviving Campbell's Paradigm for Attitude Research. *Personality and Social Psychology Review*, *14*(4), 351-367.

Klein, P.S., Nir-Gal, O., & Darom, E. (2000). The Use of Computers in Kindergarten, With or Without Adult Mediation; Effects on Children's Cognitive Performance and Behavior. *Computers in Human Behavior*, *16*(6), 591-608.

Koerner, A.F. & Fitzpatrick, M.A. (2004). Communication in Intact Families. In A.L. Vangelisti (ed.), *Handbook of Family Communication* (pp. 177-195). London: Routledge.

Kutner, L.A., Olson, C.K., Warner, D.E., & Hertzog, S.M. (2008). Parents' and Sons' Perspectives on Video Game Play: A Qualitative Study. *Journal of Adolescent Research*, *23*(1), 76-96.

Linacre, J.M. (2006). Facets, Winsteps. Retrieved 29 April, 2010, from http://www.winsteps.com/.

McLeod, J.M. & Chaffee, S. (1972). The Construction of Social Reality. In J.S. Tedeschi (ed.), *The Social Influence Processes* (pp. 50-99). Hawthorne, NY: Aldine.

Miniwatts Marketing Group. (2009). *Internet Usage in Europe. Internet World Stats*. Retrieved April 19, 2010, from http://www.Internetworldstats.com/stats4.htm.

Nathanson, A.I. (2002). The Unintended Effects of Parental Mediation of Television on Adolescents. *Media Psychology*, *4*(3), 207-230.

Nielsen Games. (2008). *Video Gamers in Europe – 2008*. Retrieved January 13, 2010, from http://www.isfe-eu.org.

Nikken, P. (2003). Parental Mediation of Children's Video Game Playing: A Similar Construct as Television Mediation. In M. Copier & J. Raessens (eds.), *Level Up Conference Proceedings: Proceedings of the 2003 Digital Games Research Association Conference*. Utrecht: DIGRA.

Nikken, P. & Jansz, J. (2006). Parental Mediation of Children's Videogame Playing: A Comparison of the Reports by Parents and Children. *Learning, Media and Technology*, *31*(2), 181-202.

Nikken, P., Jansz, J., & Schouwstra, S. (2007). Parents' Interest in Videogame Ratings and Content Descriptors in Relation to Game Mediation. *European Journal of Communication*, *22*(3), 315-336.

Oosting, W., IJsselsteijn, W.A., & de Kort, Y.A.W. (2008). *Parental Perceptions and Mediation of Children's Digital Game Play at Home: A Qualitative Study*. San Diego: CSCW2008.

Oosting, W., IJsselsteijn, W.A., & de Kort, Y.A.W. (in preparation). *Parents, Children, and Digital Games: An Explorative Study of Parental Mediation Strategies.*

Pratchett, R. (2005). *Gamers in the UK: Digital Play Digital Lifestyles*. Retrieved January 13, 2010, from http://open.bbc.co.uk/newmediaresearch/files/BBC_UK_Games_Research_2005.pdf.

Reitman, D., Rhode, P.C., Hupp, S.D.A., & Altobello, C. (2002). Development and Validation of the Parental Authority Questionnaire – Revised. *Journal of Psychopathology and Behavioral Assessment*, *24*(2), 119-127.

Rideout, V.J., Foehr, U.G., & Roberts, D.F. (2010). *Generation M^2: Media in the Lives of 8- to 18-Year-Olds*. Retrieved February 16, 2010, from http://www.kff.org/entmedia/8010.cfm.

Ritchie, L.D. & Fitzpatrick, M.A. (1990). Family Communication Patterns: Measuring Intrapersonal Perceptions of Interpersonal Relationships. *Communication Research*, *17*(4), 523-544.

Salen, K. & Zimmerman, E. (2003). *Rules of Play: Game Design Fundamentals*. Cambridge: The MIT Press.

Schmitz, C. (2009). *LimeSurvey*. Retrieved 5 June, 2011, from http://www.limesurvey.org/.

Valkenburg, P.M., Krcmar, M., Peeters, A.L., & Marseille, N.M. (1999). Developing a Scale to Assess Three Styles of Television Mediation: "Instructive Mediation", "Restrictive Mediation", and "Social Coviewing". *Journal of Broadcasting & Electronic Media*, *43*(1), 52-66.

Warren, R. (2003). Parental Mediation of Preschool Children's Television Viewing. *Journal of Broadcasting & Electronic Media*, *47*(3), 394-417.

Warren, R., Gerke, P., & Kelly, M.A. (2002). Is There Enough Time on the Clock? Parental Involvement and Mediation of Children's Television Viewing. *Journal of Broadcasting & Electronic Media*, *46*(1), 87-111.

PART III

CONTACT

PEER RELATIONSHIP-BUILDING AND NEW FORMS OF PEER AGGRESSION

Cyberbullying Related Behaviour and Third-Person Effect

Bernhard GOODWIN & Nayla FAWZI[1]

Public Acts of Cyberbullying

Web 2.0 provides adolescents with numerous opportunities to communicate with friends or to present themselves in a social network. But more and more they experience risky online contacts like cyberbullying on the web too. Bullying does not only take place in schools anymore, but also via social communities, chats, blogs or mobile phones. We define cyberbullying as the use of Internet and mobile phone applications to defame, compromise or inflict damage on other people. This may occur in written form, via mobile phone calls, pictures or videos (Fawzi, 2008). Typical cyberbullying acts are denigration, harassment, flaming or exclusion (Willard, 2007). In Germany every fifth teenager has been cyberbullied at least once (Grimm, Rhein, & Clausen-Muradian, 2008; MPFS, 2008). International studies report values between seven and 43 per cent of youth having been victims of cyberbullying (Hinduja & Patchin, 2008; NCH, 2005; Smith *et al.*, 2008; Ybarra, 2004; Ybarra *et al.*, 2007).

Due to the use of the Internet, important differences to traditional bullying emerge: Perpetrators and victims are invisible to each other, so perpetrators can bully anonymously. Cyberbullying can take place 24/7 and is independent from a location, so perpetrators and victims do not have to be at the same place (Webster, 2008; Willard, 2007). Furthermore, cyberbullying incidents are stored permanently and the victim cannot know who has seen it already and who has not. Cyberbullying can thus have a much wider range than traditional bullying (Fawzi, 2008; Hinduja & Patchin, 2008; Keith & Martin, 2005). This, in particular, holds true for the public cyberbullying incidents that we will focus on. In contrast to current research that takes an educational or social

1 Acknowledgements: The present study was supported during concept, data collection and first analysis by Miriam Schnick, Florian Tippelt, Nathalie Hasler, Claudia Kettenhofen, Julia Kurz and Stefanie Lindner. It would not have been possible without the support of VZnet Netzwerke Ltd, spickmich GmbH and lokalisten media GmbH with their respective staff and communities.

psychological point of view, we will analyse cyberbullying from a communication studies perspective. In our opinion these incidents are particularly interesting, because, with the emergence of Internet-based communication, private communication becomes public and can theoretically be perceived by every Internet user worldwide. Thus the victim becomes an object of "negative media coverage". Perpetrators produce media content and by doing so, they trigger media effects on the victims. In this process the observers or bystanders play a major role, because they represent the public. The present study investigates cyberbullying not only from the victims' and perpetrators' perspective but also from the observers' point of view. We will first analyse how adolescents feel emotionally when observing cyberbullying and second how these emotions influence their reaction to it.

We assume that youth involved in cyberbullying is aware of the public and that this has an effect on how such behaviour is perceived by perpetrators, victims and observers and how they react to it. Communication studies have shown that individuals estimate media effects on other people higher than on themselves. This phenomenon is called the third-person effect and will be discussed in the next paragraph. Against this theoretical background, the hypothesis of this study and its method are presented. The last two paragraphs present the results and discuss them.

Third-Person Perceptions and Their Influence on Behaviour

The third-person effect was first described by Phillips Davison. He stated that:

> [this] hypothesis predicts that people will tend to overestimate the influence that mass communications have on the attitudes and behaviour of others. [...] And whether or not these individuals are among the *ostensible* audience for the message, the impact that they expect this communication to have on others may lead them to take some action [italics in original] (Davison, 1983, p. 3).

With this hypothesis Davison follows concepts like the Thomas-theorem (Thomas & Thomas, 1928) and that of reciprocal effects (Lang & Lang, 1953; Kepplinger, 2007) – without directly referring to them. In the scope of our study, the act of cyberbullying is the perceived communication to which a certain impact is attributed and the audience includes those who perceive it. The third-person framework has been used before to assess its influence on the evaluation of defamatory content (Cohen, Mutz, Price, & Gunther, 1988).

The first part of Davison's hypothesis is commonly called the perception-component of the third-person effect. Single studies as well as meta-studies successfully confirmed it at various times and under different conditions (Davison, 1996; Perloff, 1999; Brosius & Engel, 1996; Huck & Brosius, 2007). Among other factors, the desirability of a message plays an important role in explaining the extent and the direction of third-person perceptions. While undesirable message effects are more likely to be attributed to the third person, desirable message effects reverse third-person perception (Perloff, 1993). This phenomenon has been called reversed third-person-perception or first-person perception. We will use the former term. It has been analysed in the context of public service announcements (PSAs) concerned with harmful behaviour amongst children (e.g. Henriksen & Flora, 1999) and adolescents (e.g. Cho & Boster, 2008). An explanation for the influence of desirability on the third-person-perception is sought in concepts like optimistic bias (Gunther & Mundy, 1993) and self-enhancement (Meirick, 2005). In both concepts individuals assess an undesirable effect on themselves as less probable than on others, because they think harmful effects are less likely to happen to them (optimistic bias) or because they think they are better in avoiding negative influence than others (self-enhancement). In the context of this analysis negative feelings towards incidents of cyberbullying are desirable effects, because most adolescents consider cyberbullying as anti-social behaviour.

The second part of the third-person hypothesis, the behaviour-component, has received less attention by researchers. As possible behaviour triggered by the third-person-perception most studies examined the willingness to censor such things as negative media content, however with inconclusive results (Perloff, 2002). In the context of cyberbullying a whole range of different behaviour can be analysed. This includes reactions of the victim, further acts of cyberbullying by perpetrators and supportive or aversive behaviour by observers. In this study we will concentrate on the latter.

Albert Gunther and Douglas Storey proposed a broader context of the third-person effect: the influence of presumed influence (2003). Unlike earlier research they did not concentrate on the difference between presumed influence on self and the presumed influence on others but put the focus just on the latter. Ye Sun and her colleagues built their framework on possible behavioural consequences of perceived influence on others (Sun, Shen, & Pan, 2008). They proposed different kinds of behavioural modes: accommodating and rectifying behaviour. An example in the context of cyberbullying would be an observer joining a hate-group on a social community website (accommodating behaviour) or an observer notifying the authorities about a malicious website (rectifying behaviour).

Hypotheses

This study investigates hypotheses concerning adolescents' own feelings towards cyberbullying, their perception of differences to their peers' feelings and the influence of these perceptions on their reactions after a cyberbullying incident.

First, adolescents' perception of their own feelings towards incidents of cyberbullying will be examined. Since cyberbullying conflicts with social norms, it can be assumed that adolescents perceive their own feelings towards this kind of behaviour to be negative rather than positive. Therefore Hypothesis 1 states: In the perception of their own feelings towards incidents of cyberbullying, adolescents report negative feelings more frequently than positive feelings.

Secondly, the perception of the respondents' peers is explored. How do they think other adolescents will feel towards incidents of cyberbullying? Because of optimistic bias and self-enhancement it can be assumed that they will perceive their own feelings to be more negative towards cyberbullying, than their peers' responses. In other words: Respondents will experience a reversed third-person-perception with negative feelings and a third-person perception with positive feelings towards incidents of cyberbullying (Hypothesis 2).

Within different subgroups of the respondents, different perceptions of emotional reactions towards cyberbullying can be expected. Sex, age and education will probably play a role in these perceptions, because perceptions of social norms differ within these groups. Hypothesis 3 states therefore: Respondents of different sex, age and education differ in their perceptions of own and others' feelings towards incidents of cyberbullying.

Respondents who report having been involved in cyberbullying as victims or bullies have personal experience with such incidents. Their perceptions of own and others' feelings towards such incidents will differ from other respondents. Respectively, victims will report their own negative feelings more often and show a stronger (reversed) third-person-perception than other respondents (Hypothesis 4a). Bullies will report their own negative feelings less and show a weaker (reversed) third-person-perception than other respondents (Hypothesis 4b). Adolescents who are only observers of cyberbullying will not share this strong emotional experience. Observers will not differ in their perceptions of own and others' feeling from other respondents (Hypothesis 4c).

The original third-person hypothesis stated that the differences in perceptions of effects on self and others would lead to certain behaviour. Hypothesis 5 states: The perception of own and others' feelings leads to differences in reported reactions towards an incident of cyberbullying.

Method

The aim of this study is to further the understanding of cyberbullying-behaviour by employing the third-person effect concept and the influence of presumed influence. For data collection we conducted an online survey among 14 to 20 year old German-speaking youth. One can assume that students younger than 14 also observe cyberbullying, but they were not specifically targeted and were evicted from the sample due to ethical considerations.

The survey was advertised on three popular German social network sites, with *schülerVZ* representing 41 per cent of the participants, *spick-mich* 36 per cent and *lokalisten* 20 per cent. For 3 per cent of the participants, the social network site could not be identified. The survey was conducted in January 2010. The sample reflects the exploratory nature of this study. Even though it is not a random sample, correlations between different variables can be interpreted meaningfully. The fact that 97 per cent of German youth are online (MPFS, 2008) counters the objection that only those with a connection to the Internet could take part. Furthermore it is obvious that the probability of coming into contact with cyberbullying is much higher when using the Internet.

Besides questions about demographics we asked participants whether they have experienced cyberbullying as a bully, victim or observer. Subsequently they were asked how they would feel if they were to witness cyberbullying as compared to other youth witnessing such an incident (given feelings had to be rated on a 5-item scale). The behavioural component was then measured additionally in terms of prospective behaviour (given behaviour had to be rated on a 5-item scale).

2,502 teenagers from 14 to 20 years old filled in the questionnaire – respondents who were not within this age-range were removed from the sample. The average age was 16.1 years – with 45.8 per cent younger than 16 and 54.2 per cent 16 and older. 54 per cent of the respondents were male, 46 per cent female. Like most online surveys among youth, academic high schools students (German *Gymnasium*) were overrepresented with 52.4 per cent followed by secondary modern school students (German *Realschule*) (20.1 per cent). All together 27.5 per cent attended another type of school (e.g. German *Hauptschule*, *Berufsschule*, *Gesamtschule*) or had finished school already. 28.3 per cent of the respondents reported they had been victims of cyberbullying (N = 2,481), 23.6 per cent admitted being part of acts of cyberbullying against others (N = 2,473). 63.2 per cent of the respondents said they had observed incidents of cyberbullying (N = 2,461).

Due to the large sample, almost all statistical tests were computed as significant. Therefore we do not report the results of null-hypothesis significance testing, but those of effect-tests and equivalence-tests both with a small effect size of $\delta = 0.2$ or $\psi = 0.1$ (Levine, Weber, Park, & Hullett, 2008; Cohen, 1988; Klemmert, 2004). This means that significant differences in this report are greater (with the given probability) than a difference which could be considered a small effect and that significant equivalence is reported, if the difference (with the given probability) is smaller than a difference which could be considered a small effect.

Results

To test the first hypothesis, respondents were presented with a fictional situation, where they came across a hate-group in a social community, in which other adolescents were engaged in an activity of cyberbullying. The fictional victim of the described situation is an adolescent to whom they have a neutral position. The respondents were asked how they would feel in such a situation. Seven different emotions were given and the respondents could rate on a scale of five how strong they would experience such an emotion, with low values representing weaker perceptions of the emotion and higher values stronger perceptions. Five of them were negative emotions (*angry*, *sad*, *helpless*, *insecure* and *afraid*), the other two were non-negative emotions (*neutral* and *amused*).

Table 1: Own feelings after observing cyberbullying within an online-group

	statistics			means in subgroups[1]							
	mean	st-dev	N	male	fem.	14-15	16-20	Gym	vict	bully	obs
angry	3.43	1.34	2371	3.16	3.73	3.39	3.46	3.48	3.57	3.19	3.44
				***		ooo		oo			ooo
sad	2.79	1.38	2355	2.48	3.14	2.70	2.87	2.91	2.87	2.56	2.76
				***		o			oo		oo
Helpless	2.21	1.31	2351	1.93	2.54	2.21	2.22	2.23	2.33	2.10	2.16
				***		ooo		ooo		o	o
Insecure	2.28	1.31	2376	2.06	2.53	2.35	2.23	2.34	2.40	2.21	2.26
				***		oo		oo		oo	ooo
afraid	1.68	1.04	2346	1.48	1.91	1.72	1.65	1.69	1.77	1.61	1.67
				***		ooo		ooo		o	ooo
Neutral	2.17	1.33	2285	2.36	1.96	2.21	2.13	2.14	2.13	2.55	2.22
				**		ooo		ooo	ooo	***	o
Amused	1.69	1.07	2367	1.84	1.53	1.68	1.70	1.70	1.69	2.19	1.78
				*		ooo		ooo	ooo	***	

[1] In subgroups where only one mean is given (*Gymnasium*, *victim*, *bully* and *observer*) the t-test was calculated among members vs. non-members of the subgroup.

Significance of a small effect between groups $p_{(d > 0.2)}$: *** $p < 0.1\%$; ** $p < 1\%$; * $p < 5\%$;
Significance of equivalence between groups $p_{(d < 0.2)}$: °°° $p < 0.1\%$; °° $p < 1\%$; ° $p < 5\%$.

Table 1 gives the results for all respondents and for those in particular subgroups. Hypothesis 1 states: In the perception of own feelings towards incidents of cyberbullying adolescents report negative feelings more frequently than positive feelings. This is especially true for the feelings *angry* (49.5%),[2] *sad* (32.1%) and *amused* (8%). For these emotions the hypothesis is supported. For other emotions like *insecure* (19.7%), *helpless* (17.7%) and *neutral* (16.1%) the hypothesis is not that strongly supported. The fact that the respondents supported the emotion *afraid* (6.8%), a negative emotion least frequently, contradicts the first hypothesis. In summary, hypothesis 1 is almost fully supported by the data of this study.

To test the third-person-perception of respondents, they were asked to report their perception on their peers' emotions after observing the described incident of cyberbullying. To calculate the strength of a third-person-perception, the differences between the perceived feelings of their peers and their own reported feelings were computed. In this kind of calculation, positive values are indicative for a third-person perception and negative values show a reversed third-person perception.

Table 2: Third-person perception of observing cyberbullying within an online group

	Statistics			means in subgroups[1]							
	mean	**st-dev**	**N**	**male**	**fem.**	**14-15**	**16-20**	**Gym.**	**vict**	**bully**	**obs**
angry	-0.60	1.43	1936	-0.42	-0.79	-0.58	-0.62	-0.67	-0.74	-0.48	-0.66
						°°°		°		°	°
sad	-0.43	1.52	1904	-0.21	-0.67	-0.36	-0.50	-0.57	-0.54	-0.23	-0.40
				*		°			°		°°
helpless	0.27	1.59	1905	0.43	0.09	0.27	0.27	0.27	0.23	0.30	0.33
						°°°		°°°	°°°	°°°	°
insecure	0.32	1.56	1925	0.45	0.17	0.24	0.39	0.28	0.26	0.31	0.36
						°		°°°	°°	°°°	°°
afraid	0.47	1.39	1901	0.58	0.34	0.41	0.51	0.44	0.45	0.53	0.50
						°°		°°°	°°°	°°	°°
Neutral	0.99	1.69	1859	0.75	1.23	0.82	1.13	1.14	1.19	0.78	1.02
				*							°°°
Amused	1.42	1.56	1960	1.24	1.59	1.36	1.46	1.44	1.63	1.17	1.44
						°°		°°°			°°°

[1] In subgroups where only one mean is given the (Gymnasium, victim, bully and observer) t-test was calculated among members vs. non-members of the subgroup.
Significance of a small effect between groups p (d > 0.2): *** p < 0.1%; ** p < 1%; * p < 5%;
Significance of equivalence between groups p (d < 0.2): °°° p < 0.1%; °° p < 1%; ° p < 5%;

Table 2 shows the results of these computations for all respondents as well as for particular subgroups. Hypothesis 2 stated: Respondents will

2 The following values of percentage refer to adolescents who rather or fully agreed with the emotions.

experience a reversed third-person-perception with negative feelings and a third-person-perception with positive feelings towards incidents of cyberbullying. This hypothesis is supported by the fact that *angry* and *sad* have a negative sign and therefore are instances of a reversed third-person-perception. Also *amused* and *neutral* are clear examples of a third-person-perception. The hypothesis is not supported for the emotions *helpless*, *insecure* and *afraid.* In summary, the data do not fully support hypothesis 2.

Hypothesis 3 stated: Respondents of different sex, age and education differ in their perceptions of own and others' feelings towards incidents of cyberbullying. This hypothesis was tested with the seven feelings the respondents reported. Table 1 and table 2 report the facts for the different groups. Regarding the feelings of the respondents themselves, only the male and female adolescents differ significantly (i.e. with a measurable effect greater than a small effect of $d = 0.2$) with female respondents rating higher on negative feelings and lower on positive feelings than male respondents. The average ratings among the two different age groups (14-15 vs. 16-20) were statistically equivalent. This also holds true for all emotions but *sad*, if students with higher formal education (*Gymnasium*) are compared with all other respondents.

Testing the hypothesis for third-person-perceptions of respondents, only two effects are significantly greater than a small effect. In both cases (*sad* and *neutral*) a larger (reversed) third-person-perception can be observed amongst female participants. Almost all differences between the age groups and the respondents with higher vs. lower formal education are statistically equivalent, except *neutral* in both comparisons, and *sad* when comparing levels of formal education.

To investigate the respondents' perceived own and others' emotions, two factor analyses were conducted with the recipients' own feelings and the third-person-perception. The data was appropriate for factor analysis with KMO 0.75 and 0.77 and significant Bartlett tests ($p < 0.1\%$). Two factors with eigenvalues greater than 1 could be extracted. The first factor in both analyses will be called empathy with victims and the second disgust towards cyberbullying, the latter having been reversed from the original result of the factor analysis in the case of own feelings for better interpretation. Table 3 shows the results of the analysis, which have been adjusted in the described way.

Table 3: Factor analysis of perceived feelings towards incidents of cyberbullying

	perceived own feelings		third-person-perception	
	empathy	**disgust**	**empathy**	**disgust**
angry	0.19	**0.73**	0.17	**0.73**
sad	0.54	0.51	0.46	0.61
helpless	**0.81**	0.10	**0.82**	0.14
insecure	**0.84**	0.04	**0.84**	0.08
afraid	**0.87**	0.04	**0.87**	0.09
neutral	-0.02	**-0.81**	0.02	**-0.78**
amused	0.02	**-0.71**	-0.07	**-0.78**
% explained variance	34.93	27.98	33.88	30.81
	62.90		64.69	

Table 4 helps to continue the testing of hypothesis 3: Again, the differences among the respondents of different sex are significantly larger than a small effect, except for the third-person-perception of empathy with the victims. The female adolescents show more own empathy and disgust, and a larger reversed third-person perception of disgust towards cyberbullying. Both age and education have no significant effect on own feelings and on the third-person-perception of empathy. In summary, hypothesis 3 can only be supported for the differences in sex of the respondents. On the basis of the data presented in this report, it has to be rejected regarding the differences in age and formal education.

Hypothesis 4a states: Victims will report more own negative feelings and a stronger (reversed) third-person-perception than other respondents. The data presented in table 1 shows that statistically victims have the same positive feelings towards incidents of cyberbullying and there is neither equivalent (except in the case of *sad*) nor a measurable effect on different negative feelings. Table 2 shows that there are no significant effects on the (reversed) third-person-perception, but a statistical equivalence in the case of the feelings *sad*, *helpless*, *afraid* and *insecure*. Table 4 shows that own disgust and third-person-perception of empathy are statistically equivalent and there is no significant effect among the other dimensions of emotions can be shown. In summary hypothesis 4a has to be rejected.

Table 4: Empathy with victims and disgust towards cyberbullying

	Statistics[1]	means in subgroups[2]							
	N	male	fem	14-15	16-20	Gym	vict	bully	obs
own feelings: empathy	2112	-0.23	0.26	0.03	-0.02	0.03	0.09	-0.04	-0.02
		***		°°°		°°		°°	°°°
own feelings: disgust	2112	-0.19	0.21	-0.05	0.05	0.04	0.04	-0.41	-0.05
		***		°		°°	°°	***	
third-person-perc.: empathy	1591	0.07	-0.08	-0.06	0.05	-0.01	0.00	-0.01	0.04
				°		°°°	°°°	°°°	
third-person-perc.: disgust	1591	0.17	-0.16	0.08	-0.07	-0.08	-0.16	0.15	-0.03
		**							°°

[1] The reported values are standardised with mean=0 and standard deviation=1.
[2] In subgroups where only one mean is given (Gymnasium, victim, bully and observer) the t-test was calculated among members vs. non-members of the subgroup.
Significance of a small effect between groups p (d > 0.2): *** p < 0.1%; ** p < 1%; * p < 5%;
Significance of equivalence between groups p (d < 0.2): °°° p < 0.1%; °° p < 1%; ° p < 5%;

Hypothesis 4b states: Bullies will report less own negative feelings and a weaker (reversed) third-person perception than other respondents. Table 1 shows that bullies report significantly stronger non-negative emotions (*neutral* and *amused*), equivalent feelings *helpless*, *afraid* and *insecure* as well as statistically inconclusive differences in *angry* and *sad*. The data presented in table 2 shows no significant effects on the (reversed) third-person-perception but equivalence in the emotions *angry*, *helpless*, *afraid* and *insecure*.

From table 4 it can be read, that bullies report significantly less own disgust towards cyberbullying and the same degree of own feeling and third person-perception of empathy for victims of cyberbullying. In summary hypothesis 4b has to be rejected almost fully, with the exception of the expression of own disgust towards cyberbullying which is weaker among adolescents who report to have been part of cyberbullying-attacks.

Hypothesis 4c states: Observers will not differ in their perceptions of own and others' feeling from other respondents. The data in table 1 and table 2 shows, that the reported own emotions (with the exception of *amused*) and the third-person-perception do not differ. As table 4 shows, this fact does not hold true for all calculated dimensions of emotions. While the reported own empathy and the third-person-perception of disgust are equivalent, the differences among the other dimensions are inconclusive. In summary hypothesis 4c is supported almost fully with the exception of some inconclusive findings regarding the own expression of disgust towards cyberbullying and the third-person-perception of empathy.

We will now turn to the possible behavioural consequences of own and others' feelings about cyberbullying. A descriptive analysis of the reported adolescents' reactions after fictionally observing cyberbullying is presented in table 5. It shows that respondents say they would first report the group to the provider, support the victim and ask the perpetrators to stop bullying. Fewer respondents said they would do nothing or contact an adult like a teacher or a parent. Only some adolescents said they would join the group.

Female respondents are clearly more prepared to contact an adult than male respondents. They are only equivalent regarding the possibility to do nothing after observing an incident of cyberbullying. Respondents in the different age groups show no differences in the way they say they would react to cyberbullying. The same is true for all reactions except for *join group*, if pupils with higher formal education (*gymnasium*) are compared with the other respondents.

Table 5: Reactions to observing cyberbullying in an online-group

	statistics			means in subgroups[1]							
	mean	**st-dev**	**N**	**male**	**fem.**	**14-15**	**16-20**	**Gym**	**vict**	**bully**	**obs**
report	4.04	1.36	2430	3.90	4.20	3.99	4.08	4.07	4.08	3.56	3.97
group						°°°		°°°	°°°	***	
support	3.13	1.27	2347	2.98	3.30	3.15	3.12	3.11	3.23	2.83	3.12
victim						°°°		°°°	°	**	°°°
tell bull. to	2.99	1.41	2365	2.87	3.11	3.05	2.93	2.97	3.07	2.78	2.99
stop						°°		°°°	°°		°°°
contact	2.16	1.38	2379	1.93	2.41	2.17	2.15	2.15	2.18	1.78	2.06
adult				***		°°°		°°°	°°°	***	
do nothing	2.60	1.42	2378	2.65	2.55	2.69	2.53	2.61	2.61	2.87	2.66
				°°		°		°°°	°°°		°
join group	1.37	0.89	2414	1.46	1.27	1.42	1.33	1.29	1.40	1.79	1.45
						°°			°°°	***	

[1] In subgroups where only one mean is given (Gymnasium, victim, bully and observer) the t-test was calculated among members vs. non-members of the subgroup.
Significance of a small effect between groups p (d > 0.2): *** p < 0.1%; ** p < 1%; * p < 5%;
Significance of equivalence between groups p (d < 0.2): °°° p < 0.1%; °° p < 1%; ° p < 5%;

Personal experience with cyberbullying has some influence on the reported reactions of the adolescents. Bullies' reactions differ significantly from the reactions of other respondents. They are less likely to report the group, support the victim or contact an adult, but are more likely to join the hate-group. Victims do not differ significantly in their reactions towards incidents of cyberbullying, while the differences between respondents who had observed such incidents before and other respondents are either inconclusive (*report group, contact adult* and *join group*) or do not exist statistically (*support victim, tell bullies to stop* and *do nothing*).

Hypothesis 5 states: The perception of own and others' feelings leads to differences in reported reactions towards an incident of cyberbullying.

To test these hypotheses two kinds of regression-analysis were computed. The findings of these analyses are presented in table 6 with the dimensions of own emotions as key-predictors – and in table 7 with the dimensions of third-person-perceptions as key-predictors – of the reactions to the fictional incident presented to the respondents in the survey. In both cases the other predictors were *age* in years, *sex* (1 for male and 2 for female respondents) and *education* (0 for not in the higher formal education institution *Gymnasium* and 1 for *Gymnasium*). Additionally the fact whether respondents said they had been *victims* of cyberbullying before, or *bullies* themselves or *observers* of such incidents itself was included (with 0 for not being and 1 for being part of this group).

Analysing the influence of own feelings towards a fictional incident of cyberbullying on reported reactions to such an event (table 6) reveals a clear influence of the dimension disgust towards cyberbullying. Adolescents, who report high values of disgust, are more likely to report such groups to a provider, support the victim of the incident, tell the bullies to stop or contact an adult, while they are less likely to do nothing or join the hate-group. On the other hand the dimension empathy with the victim has no significant effect on the reactions after an incident of cyberbullying, just the effect on the willingness to contact an adult is inconclusive. Age, sex and education do not contribute to the possible reactions statistically. Personal experience statistically does not contribute either, except for the influence of the fact that a respondent had been involved in a cyberbullying-attack as a bully on the reaction to join the hate group (which is inconclusive).

Table 6: Own feelings as predictors of reactions observing cyberbullying

	age	sex	edu.	vict.	bully	obs.	emp	disg	R^2_{corr}
report group	0.02 °°°	-0.01 °°°	0.00 °°°	0.05 °°°	-0.10 °°°	-0.02 °°°	0.10 °°°	0.44 ***	0.24 ***
support victim	-0.03 °°°	0.00 °°°	-0.04 °°°	0.04 °°°	-0.05 °°°	0.01 °°°	0.10 °°°	0.47 ***	0.24 ***
tell bull. to stop	-0.06 °°°	-0.02 °°°	-0.04 °°°	0.00 °°°	0.00 °°°	0.04 °°°	0.06 °°°	0.43 ***	0.18 ***
contact adult	-0.02 °°°	0.06 °°°	-0.03 °°°	0.03 °°°	-0.07 °°°	-0.05 °°°	0.18	0.33 ***	0.18 ***
do nothing	-0.04 °°°	0.03 °°°	0.02 °°°	0.00 °°°	0.00 °°°	0.02 °°°	0.09 °°°	-0.47 ***	0.22 ***
join group	-0.05 °°°	-0.05 °°°	-0.08 °°°	-0.07 °°°	0.20	0.05 °°°	0.01 °°°	-0.29 ***	0.18 ***

N = 2032; Standardised beta-coefficients of the regression and the significance of a t-test (with a small effect of $\beta = 0.2$), besides column R^2corr, which reports the corrected R^2-value of the regression and the significance of an F-Test (with a small effect of $\psi = 0.1$).

Effect: *** $p < 0.1\%$; ** $p < 1\%$; * $p < 5\%$; Equivalence: °°° $p < 0.1\%$; °° $p < 1\%$; ° $p < 5\%$;

The regressions explain between 18 and 24 per cent of the variance in the reported reactions on an incident of cyberbullying. These explanations are significantly stronger than small effects and can be viewed as medium to large effects. The comparison of the results of regression-analyses with the present results without the dimensions of own emotions reveals the additional variance explained by the inclusion of the respondents' dimensions of own emotions. This additional variance is quite substantial: 18 per cent for *report group*, 20 per cent for *support victim*, 16 per cent for *tell bullies to stop*, 12 per cent for *contact adult*, 21 per cent for *do nothing* and 8 per cent for *join group*.

Table 7 shows the influence of third-person-perceptions on reported reactions to an incident of cyberbullying. Again, the dimension disgust is a good predictor. Adolescents, who report low values of a third-person-perception of disgust, are more likely to report such groups to a provider, support the victim of the incident and tell the bullies to stop, while they are less likely to do nothing. The effects on the reaction to contact an adult are statistically nonexistent, while the effect on the reaction to join the group is inconclusive. The third-person-perceptions in the dimension empathy with the victim have significantly no effect on the reactions after an incident of cyberbullying. The same is true for age, sex and education. Again, only the fact, that someone has been a bully explains some variance in certain reactions. Joining the hate-group is significantly more frequent with bullies and reporting the group is inconclusive. Other personal experiences do not contribute to the reactions statistically. These regressions explain between 7 and 12 per cent of the variance in the reported reactions, which is considerably lower than the regressions described earlier. These explanations are significantly stronger than small effects, but are not even medium effects. The additional variance explained with the inclusion of the respondents' dimensions of third-person-perceptions is smaller than the reported additional variance above: 6 per cent for *report group*, 7 per cent for *support victim*, 6 per cent for *tell bullies to stop*, 2 per cent for *contact adult*, 7 per cent for *do nothing* and 3 per cent for *join group*.

Table 7: Third-person perception as predictor of reactions observing cyberbullying

	age	sex	edu	vict	bully	obs	emp	disg	R^2_{corr}
report group	0.03 °°°	0.06 °°°	-0.01 °°°	0.08 °°°	-0.18	-0.05 °°°	0.03 °°°	-0.26 **	0.12 ***
support victim	-0.02 °°°	0.08 °°°	-0.04 °°°	0.07 °°°	-0.14 °°	-0.02 °°°	0.10 °°°	-0.25 **	0.11 ***
tell bull. to stop	-0.05 °°°	0.04 °°°	-0.04 °°°	0.03 °°°	-0.08 °°°	0.01 °°°	0.08 °°°	-0.24 *	0.08 ***
contact adult	-0.01 °°°	0.15 °	-0.02 °°°	0.07 °°°	-0.14 °°	-0.07 °°°	0.04 °°°	-0.13 °°	0.07 ***
do nothing	-0.05 °°°	-0.01 °°°	0.03 °°°	-0.02 °°°	0.07 °°°	0.05 °°°	-0.14 °	0.23 *	0.09 ***
join group	-0.06 °°°	-0.07 °°°	-0.07 °°°	-0.08 °°°	0.25 *	0.06 °°°	-0.01 °°°	0.17	0.12 ***

N = 1421; Standardised beta-coefficients of the regression and the significance of a t-test (with a small effect of $\beta = 0.2$), besides column R^2corr, which reports the corrected R^2-value of the regression and the significance of an F-Test (with a small effect of $\psi = 0.1$).

Effect: *** $p < 0.1\%$; ** $p < 1\%$; * $p < 5\%$; Equivalence: °°° $p < 0.1\%$; °° $p < 1\%$; ° $p < 5\%$.

In summary Hypothesis 5 can only be supported for the own expression of disgust towards incidents of cyberbullying with regard to all reactions and for the third-person-perception of disgust in the given reactions, except for *contact adult* and *join group*.

To conclude this paragraph we want to review the hypotheses tested in this study: The first two hypotheses are supported for most of the emotions tested. Hypothesis 3 can only be supported for the differences in sex of the respondents. Hypothesis 4a has to be rejected. The same goes for Hypothesis 4b, with the exception of the expression of bullies weaker disgust towards cyberbullying. Hypothesis 4c is supported almost fully. Hypothesis 5 can be supported fully for the expression of own feelings and partly for the third-person-perception of disgust towards incidents of cyberbullying.

Discussion

In this paragraph we discuss the main findings of our study and consider some implications for further research. First, there is a relationship between the third-person-perception of cyberbullying and intended reactions. This offers some support for the behavioural component of the third-person effect hypothesis. The limitations of this finding are that no real behaviour was tested, but only behavioural intentions, and that perceived own emotions were better predictors of these behavioural intentions than the third-person-perception of these emotions. Nevertheless, these findings can help to further the understanding of the role the public plays as an additional factor in cyberbullying-attacks.

Second, disgust toward cyberbullying is a better predictor of the reported reactions to a fictional incident of cyberbullying than empathy with the victim of such an incident. This finding can help to prepare more efficient campaigns against cyberbullying targeted at adolescents. To support anti-cyberbullying rectifying behaviour of adolescents, it is better to inspire emotions of disgust toward cyberbullying than to call for empathy with the victim. This includes discouraging feelings like amusement about incidents of cyberbullying.

Third, there is a link between adolescents' sex and their own emotions about and third-person-perceptions of cyberbullying. This calls for further research to investigate whether different norms about cyberbullying among male and female adolescents lead to this effect, or whether other reasons like different perceptions of such incidents explain these facts. Scharrer and Leone (2008) have shown the influence of norms on third-person-perception in the context of video games. For practical implications the findings suggest that campaigns against cyberbullying for female and male adolescents may have to differ, since they have different perceptions of the behaviour which these campaigns want to discourage.

This study focussed on the reactions of observers of an incident of cyberbullying. This is important because this group plays a crucial role in providing possible support to the victims but also as a public for the perpetrators which might be interpreted as silent support. But the effects of third-person perception on possible behaviour should also be investigated from the point of view of victims and perpetrators.

In addition, the account of self efficacy as a predictor of either accommodating (lower self efficacy) or rectifying behaviour (higher self efficacy) should be analysed. Francis Lee suggests "a possible relationship between self-efficacy and prevention behaviour: When people are less confident about their own capabilities, they should be more concerned with the potential negative outcome" (Lee, 2010, p. 91). Such a relationship could help to explain why adolescents react differently to cyberbullying.

Additionally a way to test the causal direction of the relationship between third-person-perception and intended reactions has to be found. It might be possible to artificially manipulate the third-person-perception through the influence of exemplification (Brosius & Bathelt, 1994; Daschmann, 2000). This would mean that test subjects would be provided with different indirect information about their peers' emotions towards cyberbullying and different levels of (reversed) third-person-perception could therefore be produced.

It could be shown that public acts of cyberbullying and their perception can be better understood using the concept of the third-person effect. Understanding different possible kinds of behaviour as different reactions to the perceived influence of the cyberbullying-content can help to encourage observers to take action against the cyberbullying they observe. With further research the framework of the third-person effect can also be used to empower victims to proactively counter cyberbullying and to broaden the understanding of perpetrators and thus foster campaigns to discourage such behaviour.

References

Brosius, H.-B. & Bathelt, A. (1994). The Utility of Exemplars in Persuasive Communications. *Communication Research, 21*(1), 48-78.

Brosius, H.-B. & Engel, D. (1996). The Causes of Third-Person Effects: Unrealistic Optimism, Impersonal Impact, or Generalized Negative Attitudes Towards Media Influence? *International Journal of Public Opinion Research, 8*(2), 142-162.

Cho, H. & Boster, F.J. (2008). First and Third Person Perceptions on Anti-Drug Ads Among Adolescents. *Communication Research, 35*(2), 169-189.

Cohen, J. (1988). *Statistical Power Analysis for the Behavioral Sciences.* Hillsdale: Lawrence Erlbaum Associates.

Cohen, J., Mutz, D.C., Price, V., & Gunther, A.C. (1988). Perceived Impact of Defamation: An Experiment on Third-Person Effects. *Public Opinion Quarterly, 52*(2), 161-173.

Daschmann, G. (2000). Vox Pop & Polls: The Impact of Poll Results and Voter Statements in the Media on the Perception of a Climate of Opinion. *International Journal of Public Opinion Research, 12*(2), 160-181.

Davison, W.P. (1983). The Third-Person Effect in Communication. *Public Opinion Quarterly, 47*(1), 1-15.

Davison, W.P. (1996). The Third-Person Effect Revisited. *International Journal of Public Opinion Research, 8*(2), 113-119.

Fawzi, N. (2009). *Cyber-Mobbing. Ursachen und Auswirkungen von Mobbing im Internet.* Baden-Baden: Nomos.

Grimm, P., Rhein, S., & Clausen-Muradian, E. (2008). *Gewalt im Web 2.0. Der Umgang Jugendlicher mit gewalthaltigen Inhalten und Cyber-Mobbing sowie die rechtliche Einordnung der Problematik.* Berlin: Vistas.

Gunther, A.C. & Mundy, P. (1993). Biased Optimism and the Third-Person Effect. *Journalism Quarterly, 70*(1), 58-67.

Gunther, A.C. & Storey, J.D. (2003). The Influence of Presumed Influence. *Journal of Communication, 53*(2), 199-215.

Henriksen, L. & Flora, J.A. (1999). Third-Person Perception and Children: Perceived Impact of Pro- and Anti-Smoking Ads. *Communication Research, 26*(6), 643-665.

Hinduja, S. & Patchin, J.W. (2008). Cyberbullying: An Exploratory Analysis of Factors Related to Offending and Victimization. *Deviant Behavior*, *29*(2), 129-156.

Huck, I. & Brosius, H.-B. (2007). Der Third-Person-Effekt – Über den vermuteten Einfluss der Massenmedien. *Publizistik*, *52*(3), 355-374.

Keith, S. & Martin, M.E. (2005). Cyber-Bullying: Creating a Culture of Respect in a Cyber World. *Reclaiming Children and Youth*, *13*(4), 224-228.

Kepplinger, H.M. (2007). Reciprocal Effects: Toward a Theory of Mass Media Effects on Decision Makers. *Harvard International Journal of Press/Politics*, *12*(2), 3-23.

Klemmert, H. (2004). *Äquivalenz- und Effekttests in der psychologischen Forschung*. Frankfurt am Main: Peter Lang.

Lang, K. & Lang, G.E. (1953). The Unique Perspective of Television and Its Effect: A Pilot Study. *American Sociological Review*, *18*(1), 3-12.

Lee, F.L.F. (2010). The Prevention Effect of Third-Person Perception: A Study on the Perceived and Actual Influence of Polls. *Mass Communication and Society*, *13*(1), 87-110.

Levine, T.R., Weber, R., Park, H.S., & Hullet, C.R. (2008). A Communication Researchers' Guide to Null Hypothesis Significance Testing and Alternatives. *Human Communication Research*, *34*(2), 188-209.

Medienpädagogischer Forschungsverbund Südwest (MPFS) (2008). *JIM-STUDIE 2008. Jugend, Information, (Multi-) Media. Basisuntersuchung zum Medienumgang 12- bis 19-Jähriger in Deutschland.* Stuttgart. Retrieved February 20, 2010, from http://www.mpfs.de.

Meirick, P.C. (2005). Rethinking the Target Corollary: The Effects of Social Distance, Perceived Exposure, and Perceived Predispositions on First-Person and Third-Person Perceptions. *Communication Research*, *32*(6), 822-843.

Moessner, C. (2007). Cyberbullying. *Trends & Tudes*, *6*(4), 1-4.

National Children's Home (NCH). (2005). *Putting U in the Picture*. Retrieved February 20, 2010, from http://www.filemaker.co.uk/educationcentre.

Perloff, R.M. (1993). Third-Person Effect Research 1983-1992: A Review and Synthesis. *International Journal of Public Opinion Research*, *5*(2), 167-184.

Perloff, R.M. (1999). The Third-Person Effect: A Critical Review and Synthesis. *Media Psychology*, *1*(4), 353-378.

Perloff, R.M. (2002). The Third-Person Effect. In J. Bryant & D. Zillmann (eds.), *Media Effects: Advances in Theory and Research* (2nd ed.) (pp. 489-506). Mahwah, NJ: Lawrence Erlbaum.

Scharrer, E. & Leone, R. (2008). First-Person Shooters and the Third-Person Effect. *Human Communication Research*, *34*(2), 210-233.

Smith, P.K., Mahdavi, J., Carvalho, M., Fisher, S., Russell, S., & Tippett, N. (2008). Cyberbullying: Its Nature and Impact in Secondary School Pupils. *Journal of Child Psychology and Psychiatry*, *49*(4), 376-385.

Sun, Y., Shen, L., & Pan, Z. (2008). On the Behavioral Component of the Third-Person-Effect. *Communication Research, 35*(2), 257-278.

Thomas, W.I. & Thomas, D.S. (1928). *The Child in America: Behavior Problems and Programs.* New York: Knopf.

Webster, C. (2008). *What Is Cyberbullying?* Retrieved February 20, 2010, from http://www.cyberbullying.info/whatis/whatis.php.

Willard, N.E. (2007). *Cyberbullying and Cyberthreats. Responding to the Challenge of Online Social Aggression, Threats, and Distress.* Champaign: Research Press Publishers.

Ybarra, M.L. (2004). Linkages Between Depressive Symptomatology and Internet Harassment. *CyberPsychology & Behavior*, *7*(2), 247-257.

Ybarra, M.L., Mitchell, K.J., Finkelhor, D., & Wolak, J. (2007). Internet Prevention Messages: Targeting the Right Online Behaviors. *Archives of Pediatrics & Adolescent Medicine*, *161*(2), 138-145.

Students' Perspectives on Cyberbullying Prevention Practices

Ellen M. KRAFT & Jinchang WANG

Introduction

In the 21st century youth have extended bullying beyond the traditional schoolyard into cyberspace. They are now using e-mails, instant messages, text messages, mobile phones, and social networking sites to post derogatory comments and embarrassing pictures primarily about other students and teachers (Belsey, 2006; Campbell, 2005; Shariff, 2005; Willard, 2007a). Cases appearing in the United States courts (Associated Press, 2005; Layshock v. Hermitage School District, 2006) and recent suicides (Ballou & Ellement, 2010; Halligan, 2010; Megan Meier Foundation, 2010) have called attention to schools not being well equipped to handle the problem of cyberbullying. Preventing and stopping cyberbullying is a challenge for schools. Parents, schools, educators, and legislators are searching for solutions to the cyberbullying problem.

The problem is compounded further because most cyberbullying incidents occur away from campus (Shariff, 2008; Willard, 2007a). Cyberbullying incidents that occur away from campus can have a negative impact on school climate (Feinberg & Robey, 2009; Willard, 2007b), make it difficult for victims to function in school (Feinberg & Robey, 2009), increase the risk for psychosocial problems (Ybarra & Mitchell, 2007b), and place students' safety at risk (Feinberg & Robey, 2009).

There is a fine line between what happens online away from school and what triggers in school behaviour. Students may come to school angry as a result of conflicts that occurred through online communication as there is a significant relationship "between bullying and retaliation online and at school" (Szoka &Thierer, 2009, p. 5). A student may not know what was said about him or her until he or she hears about it the next day at school (Feinberg & Robey, 2009). Schools are reluctant to discipline students as they fear being sued for violating a student's First Amendment Free Speech rights (Tully, 2007; Willard, 2007b). A school can only discipline a student for off campus speech if the speech "materially disrupts" school activities or "substantially interferes with the rights of others" (Tinker v. Des Moines, 1969, p. 509).

Prior research about cyberbullying suggests that the strategies teenagers perceived as effective are those in which they themselves can do something about the cyberbullying (Moessner, 2007; Stacey, 2008) or someone else such as the moderator of the website will stop the cyberbullying (Moessner, 2007). The strategies students perceived as ineffective are those in which adults tell them not to cyberbully (Moessner, 2007). Several cyberbullying prevention strategies have been suggested in the literature. These strategies include incorporating cyberbullying prevention into anti-bullying programs (Willard, 2007b), setting clear rules and consequences (Aftab, 2010b; Willard, 2007b), raising awareness about cyberbullying (Campbell, 2005), Internet filters (Agatston, 2007), increased supervision of students (Berson *et al.*, 2002; Campbell, 2005), acceptable use policies (Aftab, 2010b; Berson *et al.*, 2002; Willard, 2007b), Internet safety programs (Bullypolice.org, 2010; Willard, 2007b), federal legislation to criminalise cyberbullying (Megan Meier Cyberbullying Prevention Act, 2009), showing students how messages can be traced (Franek, 2006), and anonymous technological reporting of incidents for preventing cyberbullying (McCormick, 2005; Textsomeone.com, 2010).

State anti-bullying legislation has been modified to include definitions of cyberbullying and recognise it as being part of anti-bullying legislation. The *Protecting Children in the 21st Century Act* of 2008 specifies that schools teach students about "appropriate online behaviour, including interacting with other individuals on social networking websites and in chat rooms and cyber bullying awareness and response" (Protecting Children in the 21st Century Act, 2008, Sect 2(a)(3)(b) (i) & Sect 2(a)(3)(b) (ii)). In accordance with legislation schools have developed policies against cyberbullying, reporting procedures, and educational programmes (Willard, 2007b). Schools are required to monitor student Internet use in school and filter websites on school computers in order to receive federal funding from the e-rate program (Protecting Children in the 21st Century Act, 2008). Internet safety programs have been effective in increasing students' knowledge of Internet safety strategies (Chibnal, Wallace, Leicht, & Lunghofer, 2006) however; there are no research findings that this prevention strategy results in a decrease in risky Internet behaviours (Chibnall, 2006; Crombie & Trinneer, 2003).

The strategies that schools are implementing may not be the same as the strategies students perceive to be effective. In this research we surveyed 713 students to determine:

1. What do students think should happen to a teenager after being caught cyberbullying for the first time?

2. What do students think should be done to prevent cyberbullying?
3. Are cyberbullying prevention policies effective in reducing the number of victims and offenders of cyberbullying?
4. Who do teenagers perceive to be responsible for preventing and stopping cyberbullying?

Literature Review

Definition of Cyberbullying

According to Robin M. Kowalski, professor of psychology at Clemson University, "Investigators have yet to reach a consensus on how to define cyberbullying and what time parameters to impose when assessing prevalence (within a couple of months versus lifetime prevalence)" (Kowalski, 2008, p. 45). Based on Professor Kowalski's statement, a review of academic research studies about cyberbullying, and a review of research on cyberbullying by legal expert Nancy Willard, Szoka and Thierer concluded that a "precise definition (of cyberbullying) is elusive" (Szoka & Thierer, 2008, p. 4).

Cyberbullying was initially defined by Bill Belsey as "The use of information and communication technologies to support deliberate, repeated, and hostile behaviour by an individual or group that is intended to harm others" (Belsey, 2005, p. 8; Belsey, 2010, p. 1). Nancy Willard, the Director of the Centre for Safe and Responsible Internet Use, defines cyberbullying as:

> […] being cruel to others by sending or posting harmful material or engaging in other forms of social cruelty using the Internet or other digital technologies. It has various forms, including direct harassment and indirect activities that are intended to damage the reputation or interfere with the relationships of the student targeted, such as posting harmful material, impersonating the person, disseminating personal information or images, or activities that result in exclusion (Willard, 2007b, p. 1).

She gives examples of the different forms of cyberbullying:

> […] online fights with nasty or vulgar language (flaming); sending nasty mean and insulting messages (harassment); online gossip or rumours intended to damage a person's reputation (denigration); pretending to be someone else online and posting material to get that person in trouble (impersonation); sharing someone's secret or embarrassing information online (outing); tricking someone into revealing secrets or embarrassing information online (trickery), intentionally and cruelly excluding someone from an online group (exclusion); and repeated intense harassment and denigration that includes threats that create significant fear (cyberstalking) (Willard, 2007a, p. 1-2).

As new technologies become available cyberbullies are finding other ways to harass others, for instance *Happy Slapping*, a term defined as filming a violent attack using a mobile phone or video camera and posting the clip of the evidence on the web was coined in England (Saunders, 2005).

It can be difficult to determine the difference between playful teasing or joking online and cyberbullying (Shariff, 2008; Stacey, 2008). Both bullying and cyberbullying are distinguished by a power imbalance between the bully and the victim (Shariff, 2008; Stacey, 2008). In some cases cyberbullying may be accidental (Aftab, 2010b, Shariff, 2008). In other cases victims or their friends retaliate against bullies and cyberbullies online, but feel justified in the cyberbullying acts (Aftab, 2010b; Shariff, 2008; Willard, 2007a).

Cyberbullying can occur simultaneous with in school bullying, be an extension of in school bullying, or occur exclusively online (Li, 2007). A 2008 report by the Internet Safety Technical Task states that "Bullying and harassment, most often by peers, are the most frequent threats that minors face, both online and offline" (Internet Safety Technical Task Force, 2008, p. 4). The ease of access to technology, anonymity, not being able to see the victims emotions make cyberbullying an attractive alternative to bullying in the physical world (Beran & Li, 2005; Patchin & Hinduja, 2006; Shariff, 2008). Technology levels the playing field allowing people who would not necessarily bully in the physical world to bully online (Shariff, 2008). The physically small tech-savvy child is at an advantage for cyberbullying against other less tech-savvy children, but at a disadvantage for physical bullying where the children on the playground are bigger and stronger than him. A cyberbully can create a power balance online that may not be achievable in the physical world (Shariff, 2008).

Online harassment can be perpetrated by adults. Attorney Parry Aftab defines cyberbullying to be an act that occurs among minors (Aftab, 2010b). If adults are involved the incident is considered cyber-harassment or cyberstalking (Aftab, 2010b).

Prevalence of Cyberbullying

Cyberbullying is a worldwide problem (Kraft, 2006). There have been studies in the United States, Canada, Europe, and Australia to determine the prevalence of cyberbullying. The prevalence rates for cyberbullying are influenced by the measurement criteria of the researchers and when the studies were conducted (Kowalski, 2008; Kraft, 2006). The results also vary by country and range from 7% to 72%.

According to a July 2006 national study of 1,000 United States students by *Fight Crime: Invest in Kids*, one-third of adolescents and one-sixth of children reported that they have been victims of cyberbullying (Opinion Research Corporation, 2006). During January 2005, Patchin and Hinduja studied 1,500 adolescents in the United States and found that 33% of the respondents were victims of cyberbullying (Patchin & Hinduja, 2006). A more recent study in 2008 in the United States showed that victimisation was as high as 72% (Juvonen, 2008).

In 2000, the Media Awareness Network conducted a national school-based survey of 5,682 Canadian students in Grades 4 to 11. The results of that project *Young Canadians in a Wired World* (*YCWW*) revealed that "one quarter of young Internet users (25%) say that someone has e-mailed them material that said hateful things about others" (Media Awareness Network, 2001, p. 1). A more recent survey of 177 grade seven students by Canadian researcher Li found that over one quarter of the seventh grade students experienced cyberbullying (Li, 2007).

In the UK, the children's charity Action for Children (formerly the National Children's Home) surveyed 770 adolescents between the ages of 11 and 19 and found that 20% had experienced cyberbullying (NCH, 2005). The 2009 EU Kids Online II survey of 9-16-year-olds experiences of online risk in 25 European countries revealed 5% had been bullied on the Internet and 3% bullied by mobile phone or text message for a prevalence rate of 8% for electronic bullying (Livingstone, Haddon, Gorzig, & Olafson, 2009). The report stated that 19% of the participants had been bullied online or in person. The results varied by country in the European region. Cyberbullying rates were higher in countries where traditional bullying was more common (Livingstone *et al.*, 2008).

The Australian Covert Bullying Prevalence Study was commissioned by the Department of Education, Employment and Workplace Relations to understand covert forms of bullying such as cyberbullying. The data was collected from 20,832 Australian students aging between 8 to 14 years old, from over 200 schools, and 456 school staff. The data revealed that 7-10% of students in years 4 through 9 said they had been cyberbullied during the school year (Child Health Research Promotion Centre, 2009).

There is research evidence to support that cyberbullying is a growing problem in the United States and the United Kingdom. A University of New Hampshire study showed an increase in the prevalence of cyberbullying in 2005 from 2000 (Ybarra & Mitchell, 2007a). A United Kingdom longitudinal study of 2500 students ages 11 to 13 by Rivers and Noret (2009) also showed an increase in the prevalence of girls receiving at least one offensive message per term from 13% in 2002 to 15.5% in 2006 (Cowie & Colliety, 2009). United States researchers, Hinduja and Patchin, estimate that 20% of all adolescents will experience cyberbullying at some time in their lifetime (Patchin & Hinduja, 2010).

Impact of Cyberbullying

Common emotions experienced by victims of cyberbullying are anger, frustration, sadness, and fear (Beran & Li, 2005; Patchin & Hinduja, 2006; Shariff, 2008). Some victims suffer from more serious effects such as declining school performance or depression (Beran & Li, 2005; Patchin & Hinduja, 2006; Shariff, 2008). The effects of cyberbullying are more intense and serious than traditional bullying. Not knowing who the offender is, the wider audience, and 24/7 nature of online bullying intensify the effects (Beran & Li, 2005; Livingstone, 2006; Patchin & Hinduja, 2006; Shariff, 2008). Some victims have found the cyberbullying to be so intolerable that they commit suicide (Ballou & Ellement, 2010; Halligan, 2010; Megan Meier Foundation, 2010; Patchin & Hinduja, 2010). There is research by Yale University School of Medicine that found links between being a bully, being a victim of bullying, and suicide (Szoka &Thierer, 2009).

Serious Cases of Cyberbullying: Examining Parent's and School's Reactions

Cyberbullying has been given more attention as a result of recent suicides of teenagers who have been harassed by cyberbullying in the United States. There has been national media attention to serious cases of cyberbullying in which the victim committed suicide. Three cases in particular have received extensive media coverage, namely the cases of Ryan Patrick Halligan (Halligan, 2010), Megan Meier (Megan Meier Foundation, 2010, and Phoebe Prince (Ballou & Ellement, 2010). A review of commonalities among these three cases illustrates how difficult it is for parents, schools, and the legal system in the United States to stop cyberbullying.

All three victims were victims of repeated traditional bullying that had occurred during several years. Their parents were aware of the traditional bullying. Mr. Halligan had noticed that Ryan was spending more time on the Internet during the summer when the cyberbullying took place (Halligan, 2010). Megan Meier's mother was aware of the cyberbullying and Megan's underage use of MySpace.com (Megan Meier Foundation, 2010). Mr. Halligan had wanted to contact school officials, but Ryan begged him not to for fear the bullying would become worse (Halligan, 2010). Phoebe's mother actually did contact the school; however, they did not stop the bullying (Ballou & Ellement, 2010; Crimaldi, 2010). Megan Meier's mother sent her to a private school for eighth grade with the hope that her daughter would not be bullied at the new school (Megan Meier Foundation, 2010).

What was different about the parents' reaction to the cyberbullying in comparison to the traditional bullying was that they were not fully aware of the impact that the 24/7 cyberbullying was having on their son or daughter. They did not realise how unbearable the pain of cyberbullying was and that the suffering would lead their son or daughter to commit suicide. In all three cases the parents knew there was bullying, but failed to recognise how serious the situation was. They were aware that their teenager was having difficulties; the real problem was they did not know what to do to stop the bullying and cyberbullying. The bullying and cyberbullying escalated to a level where all of the victims could not take it anymore. In Phoebe's case, some school personnel were aware of the situation and even intervened, but did not know how to stop the bullying (Ballou & Ellement, 2010). Awareness about cyberbullying is increasing, however, effective measures to prevent and control the situation need to be implemented as illustrated by these three cases. The real problem in serious cases of cyberbullying was not awareness that the victim was being bullied or cyberbullied, but how to stop the bullying and cyberbullying problem so that the victim would be relieved of his/her suffering. This research focuses on finding ways to prevent cyberbullying that are effective from the students' perspective.

Current Successful Intervention Strategies

Policies and Education

The key component to a successful anti-bullying programme is to have a policy that clearly states a schools stand against bullying (Smith, Peppler, & Rigby, 2004). The policy must specify clear rules and consequences for bullying behaviour. Additionally school policies need to communicate to students the standard of online privacy that they can expect to have similar to standards of privacy that apply to students' personal property when using facilities such as lockers (Willard, 2006) The procedures that the school staff will take when bullying incidents occur should be in the policy too. Successful anti-bullying programmes also involve teacher training in building positive student teacher relationships and classroom management (Smith *et al.*, 2004). The entire school staff needs to be able to recognise bullying and cyberbullying incidents (Smith *et al.*, 2004; Willard, 2006; Willard, 2007c). Bullying and cyberbullying prevention should be incorporated into the curriculum by having classroom discussions that address bullying issues and peer relations. It is argued by Willard that the primary purpose of a policy against cyberbullying is to provide legal protection for the school (Willard, 2007c). Policies alone do not stop cyberbullying. It is necessary to educate students about the effects of cyberbullying. Some students think cyberbullying is a joke. They might stop if they knew how it really affected the victim.

Parental Involvement

Parental supervision and education can be the key to preventing cyberbullying (Aftab, 2007a; Tully, 2007). In a study by Agatston (2007) students reported that they would be more likely to report a threatening incident of cyberbullying to their parents than to an adult at school because they do not think the adults at school could help them with the cyberbullying problem (Agatston, 2007).

Research has shown that there is an incorrect perception between what teenagers actually do online and what their parents think they do (Brown, Jackson, & Cassidy, 2006). Teenagers primarily use the Internet to socialise. Parents overestimate the time that their teenager is using the Internet for school.

There is research evidence that ongoing communication about and supervision of online activities can also prevent teenagers' engaging in risk taking behaviours on the Internet. Example of risk taking behaviours are disclosing personal information, offline meetings, sharing photos, and exposure to threatening messages (Berson, Berson, & Ferron, 2002). A study by Berson *et al.* (2002) showed that when teachers, parents, or other caregivers have an ongoing dialogue about cyber-activities and monitor adolescent girls Internet use there is a "decreased tendency to engage in cyber-activities that lead to potential harm" (Berson *et al.*, 2002, p. 51). Adolescent girls who had ongoing discussions and monitoring by parents were less likely to have filled out a form that discloses personal information, agreed to meet in person with someone they met online, told personal information, or sent suggestive e-mail (Berson *et al.*, 2002). Having poorer parental monitoring and relationships with parents or other caregivers is associated with increasing frequency of victimisation online (Ybarra *et al.*, 2007a).

Technological Reporting Systems

TextSomeone is a technological reporting system that schools in the United Kingdom have found to be successful. Using the system "students can simply forward the unwanted message on to the school along with their student ID and details of when the message was received allowing them to prove the incident" (TextSomeone.com, 2010). The system has allowed for interventions that would not have taken place if the victim had to initially report the incident in person (McCormick, 2005). TextSomeone has received awards through the Education and Technology industry including the 2009 Fast Growth Business Award (TextSomeone.com, 2010).

Older Students Educating Younger Students About Cyberbullying

Johanna Mustacchi (2009) had her eighth grade students develop skits about how cyberbullying occurs, its consequences, and the ways to resolve it as part of a media literacy unit at Pierre Van Cortlandt Middle School in Croton-On-Hudson, NY (Mustacchi, 2009). The skits were presented by the eight grade students to the sixth grade students, the school psychologist, guidance counsellor, school assistance counsellor, and D.A.R.E. (drug abuse resistance) police officer. This way the eighth grade students learned about the skills needed to recognise and avoid cyberbullying and the sixth grade students were provided with valuable information (Mustacchi, 2009). The school later initiated a *cyber-citizenship* programme (Mustacchi, 2009).

Presentations by John Halligan to Raise Awareness About Cyberbullying

John Halligan travels to schools throughout the United States to tell his son Ryan's story at school assemblies and to spread the message of the impact that cyberbullying had on his son. Ryan committed suicide after relentless bullying and cyberbullying in fifth through eighth grade. He tells students in plain language about the seriousness of cyberbullying. His presentations have had a strong impact on students. Student who have bullied others have told Mr. Halligan that they decided to apologise to their victims and change their ways after hearing Ryan's story (Halligan, 2010).

Students Perspectives on Cyberbullying

What Do Students Think Cyberbullying Is?

Most research studies do not define cyberbullying based on students' perceptions of what cyberbullying is. Moreover, according to Livingstone and Haddon (2008) research using qualitative methods is used more often with younger children than teenagers (Livingstone & Haddon, 2008) and the lack of qualitative research about older teenagers "means that the findings may lack contextualisation or interpretation in terms of experiences and perception of these young people themselves" (Livingstone & Haddon, 2008, p. 317). Therefore the criteria respondents use to judge whether electronic communication is cyberbullying may not be the same as the criteria researchers use and could be a factor that causes the variation in prevalence rates among research studies (Vandebosch & Van Cleemput, 2008).

To develop a clearer definition of how students between 10 and 18 years old determine if electronic communication is considered cyberbullying, Belgium researchers Vandebosch and Van Cleemput (2008) conducted 53 focus groups from 17 classes at 10 different schools. The students were familiar with the term cyberbullying and defined it as "bullying via the Internet" (Vandebosch & Van Cleemput, 2008, p. 500). The researchers determined that "the natural descriptions given by students were similar to the ways researchers have measured cyberbullying in scientific studies" (Vandebosch & Van Cleemput, 2008), however, there was the critical difference that researchers in scientific studies did not consider the context in which the electronic communication takes place in their studies (Vandebosch & Van Cleemput, 2008). Students may consider electronic communication to be cyber-teasing or cyber-arguing (Vandebosch & Van Cleemput, 2008). Both cyber-teasing and cyber-arguing may or may not be repetitive and take place in equal power relationships (Vandebosch & Van Cleemput, 2008). Cyber-teasing is not intended to hurt whereas cyber-arguing is intended to hurt (Vandebosch & Van Cleemput, 2008). According to the findings of the study, in order to classify electronic cyber activities as cyberbullying:

> These Internet or mobile phone practices should be intended by the sender to hurt the receiver; part of a repetitive pattern of negative offline or online actions; and performed in a relationship characterized by a power imbalance (based on real life power criteria such as physical strength or age/ and/or on ICT related criteria such as technological know how and anonymity (Vandebosch & Van Cleemput, 2008, p. 502).

Hence, researchers need to take into account the context of cyber-activities when determining whether the cyberbullying activity is truly cyberbullying.

Another qualitative study by Mishna, Saini, and Solomon (2008) investigated through focus groups how Canadian students in grades 5 through 8 perceived cyberbullying. The participants in their study defined cyberbullying by comparing it to traditional bullying and identified behaviours such as spreading rumours, making threats and derogatory comments (Mishna, Saini, & Solomon, 2008). One participant in the study summarized the phenomenon as "It's another way to bully over the computer" (Mishna, *et al.*, 2008). The participants in the study had similar opinions to the findings in other research on the topic (e.g. Hinduja & Patchin, 2006; Shariff, 2008; Ybarra & Mitchell, 2004) that cyberbullying is unique because it is everywhere and anywhere and the bully can remain anonymous (Mishna, *et al.*, 2008). Some boys in the study did not perceive cyberbullying as a problem (Mishna, *et al.*, 2008).

Telling Adults

Cyberbullying is a problem that some experts believe requires adult intervention to resolve (Beasley, 2005; Aftab, 2010 b). However, students are reluctant to talk to adults about cyberbullying incidents (Campbell, 2005; Kowalski & Limber, 2007; Stacey, 2008). Teenagers believe that they should deal with cyberbullying problems by themselves (Li, 2010; Stacey, 2008). The primary reason that students do not tell adults about cyberbullying incidents is that they fear they will loose their technology privileges (Beran & Li, 2007; Keith & Martin, 2005; Mishna *et al.*, 2008; Patchin & Hinduja, 2006; Shariff, 2008). Other reasons are that adults will not understand the situation (Li, 2010; Stacey, 2008), they will not be believed (Li, 2010), the adult will not do anything (Li, 2010; Shariff, 2008), could not handle telling someone (Lines, 2007), did not think telling someone would help the situation (Li, 2010; Lines, 2007), and the situation will not change or will be come worse (Li, 2010; Lines, 2007; Stacey, 2008). Research by Li (2010) revealed that when respondents told someone (either an adult or a friend) 15% said the situation got better, 40% said it remained the same, and 6% said the situation became worse (Li, 2010).

There is a gap in communication between parents and teens about their online activities. According to a May 2009 a survey of the National Centre for Missing and Exploited Children conducted by Cox Communications, about 40% of United States teenagers tell their parents very little or nothing about their online activities, compared to only 8% telling their parents everything about what they do online (Thomas, 2009). Independent studies in the United States, United Kingdom, and Canada documented that parents and children do not discuss cyberbullying (Beran & Li, 2005; Livingstone, 2006; National Children's Home, 2005; Patchin & Hinduja, 2005). The survey of the National Centre for Missing and Exploited Children conducted by Children Cox Communications found that 77% of teenagers' parents are concerned about Internet safety and 46% of teenagers' parents have restriction on Internet use (Thomas, 2009). However, the parental controls may not be effective as 20% of the teenagers said they knew how to circumvent the controls (Thomas, 2009). An ABC News July 2009 article suggests that parents need to "get tech savvy to stop cyberbullies" (Collerton, 2009, p. 1).

Students' Perceptions of Intervention Strategies from the Research Studies

Stacey (2008) conducted focus groups with students from three different independent Australian schools asking them for suggestions for handling the cyberbullying problem. The schools were a city girls only school, a regional co-ed boarding school, and a rural co-ed school. The responses varied primarily by the age groups surveyed as well as the school that students attended. The ages and grades that were surveyed were students from the fifth to seventh grade aged 10 to 13 years old, students from the ninth grade between 13 and 15 years old, and students from the eleventh grade or 16-17-year-olds (Stacey, 2008).

The student groups surveyed identified education, technological solutions for eliminating harassers, and support systems for students as strategies that schools should consider. The education topics and who should receive the education from schools varied by the age group of respondents. The fifth to seventh grade students thought that educating both parents and students about *Internet safety* would be helpful (Stacey, 2008). The eleventh grade students thought that parents and teachers needed to be educated about social networking sites "so they don't acquire only negative impressions of social networking" (Stacey, 2008). The students of the eleventh grade thought that privacy should be discussed at school as many of them had experienced identity theft and invasion of privacy whereby their online messages had been copied and forwarded to others (Stacey, 2008).

Students also identified having support systems, but their attitudes towards different types of support systems varied by age and within an age group. Teacher intervention had mixed opinions among the younger students. Some of the fifth to seventh grade students thought teacher intervention could be helpful, but the girls at the city school did not want to talk with parents or teachers about the cyberbullying problem (Stacey, 2008). The fifth to seventh grade students were in favour of developing a "mediating peer group to help eliminate cyberbullying" (Stacey, 2008). The students from the ninth grade felt they should deal with the cyberbullying by themselves or seek support from their peers (Stacey, 2008). They varied in the support that they wanted from parents, older students, and older siblings (Stacey, 2008). The majority of the ninth grade students did not trust adults. They thought that adults do not understand the online world or that telling an adult would make the problem worse (Stacey, 2008). However, students at the boarding school had different opinions about their teachers. They did not think that the school had "a duty of care to protect them" (Stacey, 2008). The eleventh grade students would only seek help from a parent or teacher "under exceptional circumstances" (Stacey, 2008). The students from the eleventh grade thought "there should be better support systems at school" (Stacey, 2008). They suggested a "code of conduct for electronic communication" (Stacey, 2008), but viewed this intervention as being needed for "younger students rather than eleventh grade students" (Stacey, 2008). The eleventh grade students thought it was their responsibility to help younger students by sharing their online experiences with them (Stacey, 2008).

Technological interventions were suggested and used by students in the focus groups. The fifth to seventh grade students suggested technological solutions to get rid of the harasser such as "deleting a person, blocking chain mail, and tracing cyberbullies" (Stacey, 2008). The eleventh grade students at the boarding school said that having email shut down if not used properly, was helpful in protecting them against abuses of their account (Stacey, 2008).

Research about cyberbullying prevention strategies in the United States by Harris Research for the National Crime Prevention Council revealed that teenagers favoured technological solutions in which they or someone else such as the moderator of the website would stop the cyberbullying (Moessner, 2007). The teenagers surveyed did not think that strategies in which adults told them not to cyberbully were effective. They perceived the most effective strategies to be "blocking people online who bully, refusing to pass along cyberbullying messages and online groups having moderators who block offensive messages as being the most effective" (Moessner, 2007, p. 4). Strategies that were perceived as less effective were "all schools should have rules against cyberbullying; schools should educate students in small groups not to cyberbully, holding school assemblies to educate students not to cyberbully, and teaching adults to help young people not to cyberbully" (Moessner, 2007, p. 4), as these strategies were perceived as strategies in which adults tell teenagers not to cyberbully.

Kraft and Wang (2009) studied students' views on 14 cyberbullying prevention strategies. The students were grouped into four categories according to their roles played in cyberbullying: pure-offender, pure-victim, both-offender-and-victim, and neither-offender-nor-victim. Students, regardless of whether they are offenders or victims, believed that "not allowing offenders to access to social networking sites" and "parents taking away the computer and mobile phone of offenders" are among the top-three strategies that could most effectively prevent cyberbullying (Kraft & Wang, 2009). The least effective strategy rated by all the four categories of students was that cyberbullies would have to wait one year to attend a four-year college after high school and could only attend a community college (Kraft & Wang, 2009). Clear rules with penalties enforced and ongoing cyberbullying prevention programmes are in general viewed as effective by the offenders (Kraft & Wang, 2009). Offenders did not consider penalties such as not being allowed to participate in extracurricular activities; doing a presentation about cyberbullying, or attending Netiquette classes on Saturday to be effective whereas non-offenders thought these penalties would be effective. The researchers concluded that pure-offenders are deterred by the penalties restricting their access to technology rather than penalties consuming their time such as participating in community service or attending netiquette classes on Saturdays, or taking away their money, such as having to get a job and repay to the person they bullied online (Kraft & Wang, 2009).

Method

A survey instrument was used to collect the data for the study. The investigators designed the survey using the Zoomerang survey tool. The survey was approved by Richard Stockton College of New Jersey's IRB for Human Subjects committee. The data was collected from a nationwide online survey of 713 middle school and high school students from 43 states. The study was controlled to have 50% of the respondents self reporting that they were cyberbullies. The Zoomerang survey tool has a quota feature that rejects respondents based on responses after the desired sample size for a category is achieved. The limit of respondents was set to 350 for cyberbullying offenders and 350 for non-offenders based on the funding available for the study. Once the number of completed surveys reached the quota for a group, the survey was closed for that group. There were 13 extra samples for non-offenders. The 713 samples of the survey were complete and valid based on our pre-set qualifications such as ages and parents' permissions, and within the quota.

The sample was representative of the United States in terms of race, household income, and gender based on the 2000 census data. There were a total of 39 questions on the survey. Two questions were used to obtain and verify parental consent, ten were demographic questions, four were questions about students' roles as offenders and victims of cyberbullying, four were questions about their roles as bystanders to cyberbullying, fourteen were questions about their views on prevention strategies, three were questions about why teens choose to or not to cyberbully, three were questions about the cyberbullying problem and prevention measures at their school, one was a questions about how the respondent would prefer to report cyberbullying, and one about whose responsibility it was to prevent cyberbullying. Four of the 39 questions were open-ended. The respondents were asked to rate the effectiveness of a list of 10 cyberbullying prevention strategies on a scale of 1 to 5 with the following criteria: 1: very ineffective, 2: ineffective, 3: neutral, 4: effective, and 5: very effective. The other four strategies were described in detail. After reading the description of the strategy respondents were asked in a multiple choice format if they considered the strategy to be very effective, effective, neither effective nor ineffective, ineffective, or very ineffective. The four strategies presented in this matter were telling students in class what to do as a victim; setting clear rules and enforcing penalties on offender; having written policy on zero toleration about bullying; and ongoing cyberbullying prevention programmes. We presented the four strategies in this manner so that it would be more likely that students would read the descriptions carefully. The remaining questions were multiple choice questions with one answer, yes/no, or selecting multiple answers from a list.

A professional market research firm, Market Tools, Inc. was hired to recruit subjects and collect the data online. Market Tools, Inc. is a leading full-service provider of market research services. They are the parent company of www.zoomerang.com. This organisation deploys surveys created with the Zoomerang survey tool through the Zoomerang website. The people who answer the surveys are part of the ZoomPanel that consists of over two million people (Market Tools, Inc & Zoomerang.com). Market Tools e-mailed the survey to the members of the panel of ZoomPanel who indicated in their demographic profile that they have teenagers living in their home. The teenagers, who live in homes with ZoomPanel members, completed the survey for this study. The profiles of ZoomPanel members are aligned with the United States census data to ensure an accurate population profile (Market Tools, Inc & Zoomerang.com). The investigators did not have access to any personal identifiable information such as e-mail addresses. The ZoomPanel is used by multinational companies such as McDonalds, General Mills, KFC, Procter and Gamble and Microsoft (ZoomPanel.com, 2010).

The time to take the survey was estimated to be 20 minutes. The survey was administered online from 26 June 2008 to 7 July 2008.

A major limitation of the study is that the researchers never really knew who was filling in the survey. Also subjects may not be completely honest in their responses or understand the questions. The market research firm therefore estimates this survey had an error rate of 5-6%. Measures were taken by the market research firm to eliminate common markers of erroneous data. True Sample Technology verifies that participants are unique, real, and engaged (Conklin, 2009). True Sample is a unique Market Research technology that systematically ensures data quality. To ensure that data was real, Market Tools verified that information about gender, zip code, and household income matched with the validated data of consumer demographics collected in extensive databases used by the financial services industries for each member of the ZoomPanel. Market Tools used digital fingerprinting technology to check that the respondents did not repeat surveys and TrueSample technology to identify fraudulent responders and removed those records from the sample (Conklin, 2009). True Sample technology also detected common markers of erroneous data that include respondents repeating surveys, taking the survey too quickly and not reading the questions, and fraudulent data.

In order to help the participants answer the survey questions they were provided with a definition of cyberbullying and asked if they had been cyberbullied in the past year or cyberbullied others. The definition of cyberbullying was given to the respondents of the survey as follows:

Cyberbullying refers to the activities of using technology to bully others. Examples of cyberbullying were given in the survey to help students identify online bullying, such as:

1. Sending mean or threatening e-mails, instant messages, or mobile phone text messages.
2. Tricking someone into telling personal information in an e-mail. The e-mail with the personal information is forwarded it to others.
3. Creating a profile of someone on an online social networking site such as myspace.com or facebook.com without their permission.
4. Using someone else's IM screen name or e-mail account to bully others.
5. Making postings to a website to say mean things about a teenager or teacher online.
6. Creating or voting in online polls with hurtful questions such as "Who's the ugliest girl in the 9th grade".
7. Posting embarrassing pictures of another person online.
8. Making harassing or threatening phone calls with a mobile phone.

These examples were adapted from the definitions of flaming, denigration, impersonation, outing, trickery as defined by Willard (2007a). The 713 respondents self-reported whether they had been perpetrators or victims of cyberbullying by answering yes or no to each of the questions below:

1. Have you bullied anyone online in the past year?
2. Have you been a victim of cyberbullying in the past year?

Students were also asked two open ended questions: "What should happen to a teenager after being caught cyberbullying for the first time?" and "What do you think should be done to prevent cyberbullying?" They had to supply an answer to each of these questions. The questions were open ended to determine what ideas students had about preventing cyberbullying that might be different from those of the researchers. The responses were reviewed and codes were developed to categorise the responses.

To determine whether school policies were effective the students were asked: Does your school have a written policy not allowing online bullying? They were given choices of yes, no, or don't know. To determine who students thought were responsible for preventing cyberbullying, the students were asked: Whose responsibility do you think it is to have rules and penalties for teenagers to stop them from cyberbullying? Students could select more than one choice from seven stakeholders. Finally students were asked: Which way would you prefer to report a cyberbullying incident? The student chose the method from a list of choices of how they would like to report an incident of cyberbullying.

Results

What Should Happen to a Teenager After Being Caught Cyberbullying for the First Time?

Based on the responses to this question, the researchers identified common responses of the students. The results of the responses were classified as punishments; communication with the offender about the seriousness of the offence; no answer; tell the parents and have them decide what to do; do nothing; apologise; don't know or depends on the offence; arrest the offender or put him in jail or a juvenile detention centre; and community service. 48 respondents did not answer this question. The results are shown in Table 1.

Table 1: What should happen to a teenager after being caught cyberbullying for the first time?

Response	Percent
Punishment	49% (349)
Take away technology such as computer, cell phone, Internet	30% (229)
Unspecified Suspension/Punishment Grounding	9% (69)
Arrest/Jail/Juvenile Hall	4% (29)
Fines	3% (22)
Communication with offender about seriousness of the offence	
Issue written or oral warning	16% (122)
Talking to offender/reprimand/counselling	7% (56)
No answer	6% (48)
Tell parent and have them deal with offender	6% (48)
Nothing	5% (38)
Don't know/Depends	5% (32)
Apologize	2% (19)
Community Service	2% (19)
Other	5% (41)

Based on the results 72% of the students surveyed thought something should be done when a teenager is caught cyberbullying for the first time. They suggested punishment (49%); issuing an oral or written warning (16%); talking to, counselling, or reprimanding the offender (7%); telling the offender's parents about the issue (6%), having the offender apologize (2%) or having the offender do community service (2%).

Punishment was the most common response suggested by 49% of the respondents. Some respondents did not specify what the punishment should be. They simply stated: "punishment"; "he or she should be punished"; "they should be punished"; "they should be punished in some way"; "a lot of punishment"; and "have a punishment". Some of the respondents may not have specified a punishment because they thought the punishment should be based on the incident. A participant specifically said: "They should be given a punishment, either by their school or police officials, depending on how serious the bullying is". Another participant identified specific punishments based on the severity of the incident by saying:

> Well, it should depend on the seriousness of the act that the "cyberbully" committed. For example, if it is something so minute like one nasty comment and that's it. Nothing should happen, because it was a once and done occasion. Something more serious such as a threatening e-mail including curse words and promises to do harm should be reported to the police. It is the same assault as if two people were standing face to face. Of course we scuffle in everyday life, but threats are very different.

Other participants indicated that the punishment should be based on the type of cyberbullying incident by saying: "depends on degree of cyberbullying"; "it depends on what they were saying"; and "depends on what the circumstance was".

Two-thirds of the respondents who recommended specific punishments suggested that the punishment would be to revoke or suspend the offender's technology privileges. They specified technologies such as the computer, Internet, mobile phone, and the television. Respondents said: "all privileges be taken away – no cell, no Internet, no TV, etc"; "the teen should not have access to a computer"; "they should be put on cyber probation"; "kicked off the web"; "they shouldn't be allowed to use the Internet or mobile phones for a certain amount of time"; "take away the computer". Respondents did not consistently agree on how long technology privileges should be suspended or revoked. The time periods suggested, were one week, two weeks, one month, three months, six months, and one year. Some answers were: "no Internet for a week"; "loss of computer time for 2 weeks"; "have the email address or IP address suspended for 3 months"; "60 days without computer"; "computer privileges taken away for 6 months"; "no Internet for a few months"; and "they shouldn't be allowed to go on the Internet for a year".

Other suggestions for punishment were that the offender be grounded or suspended, arrested, or spend time in a juvenile hall or jail. Paying fines was also suggested with responses such as "pay a fine"; "fine"; "get fined"; and "fined heavily". The person responsible for paying the fine or who should receive the money of the fine was not always specified. Some respondents thought the fine should be paid as restitution to the victim. Comments such as "pay a fine to the child being harassed" and "monetary penalty with fine going to the victim" suggested that the victim should be financially compensated. One respondent specifically identified that "the family be fined".

The person responsible for imposing the punishment was identified by the respondents as the parents, school, police, or both parents and the school. Six percent of the respondents specifically suggested that parents should be responsible for deciding what to do with the offender. Some respondents indicated who should be doing the punishing, but not what the punishment should be. The respondents said: "they should be punished by their parents"; "some sort of disciplinary action from both their parents and school (and maybe the law)"; and "the parents, teachers, and principal should take action". Other respondents specified the punishment(s) and who should be giving it. Example responses are:

> A harsh warning that they will be banned from social networking sites and/or whatever site they were bullying people on and they should be banned by legal authorities. If they are using a phone to bully they should be grounded from the phone for a long time.

> They should have their computer taken away and their parents should be warned if it persists they will be suspended from school. Parents should have to monitor child's P.C. or mobile phone usage after the reprimand.

Respondents who said they thought the offender should be "grounded" or "suspended" did not always specify whether they thought the offender should be suspended from school, have their Internet privileges suspended, be grounded in the physical world, or be grounded by having their technology privileges revoked. Some comments were that the offender was to be "grounded from computer"; "grounded by parents or punished by them"; and "suspension from school, no computer in the home".

Almost one quarter of the respondents thought that there should be an oral or written communication with the offender about the seriousness of the offence. Warning the offender orally or in writing was suggested by 16% of the offenders. Responses indicating that a warning should be issued were: "they should be given a warning"; "get a written warning"; and "they should be given a warning to stop what they are doing". "Talking to the offender", "reprimanding the offender", or "counselling the offender" was suggested by 7% of the respondents. Several participants believed that a first time offender should be educated about the effects of what he or she did. Participants expressed that the offender needs to be "spoken to and warned or educated on the way their actions could hurt or harm someone", "need to be explained how it hurts others", and "their parents should have a talk with them about how harmful is to others and make them understand that they shouldn't do it again".

Some participants suggested offenders should have to take a class: "attend classes or meet with a therapist" and "they should be obligated to take classes on cyberbullying on the weekend".

Another participant expressed the opinion that warnings do not work by saying: "Warnings don't work. They should be banned from the site for a period of time. They should be made to publicly apologize to the person bullied."

Other participants suggested consequences that were a combination of the offender apologising to the victim, participating in intervention programmes, and/or having the offender do community service. Three example responses were:

> I think the teen should go to some type of intervention program about other people's feelings and reputations and do some type of community service.

> A warning, and they should have to take a class to educate them about the harmful things cyberbullying does to people. Also, they should have to apologize publicly and do some community service.

> I think the teen should go to some type of intervention program about other people's feelings and reputations and do some type of community service.

Five percent of the respondents said they did not know what should happen or their answer would depend on the circumstances similar to the findings. One student commented "I don't know how you'd stop it since they can change their names". Since cyberbullying can range from exchange of nasty messages to threats in the physical world relying a decision on what to do about the incident on the circumstances of the situation is logical.

There were respondents who did not think anything should be done as they said "nothing". They made comments such as:

> Nothing, its teenagers. It's what they do and that's life, get over it and don't be sensitive. Bad stuff happens.

> I think nothing, because most people don't care what other people say on the Internet, they all know theirs all kinds of shit out there.

What Students Think Should Be Done to Prevent Cyberbullying?

The responses to the question "What do you think should be done to prevent cyberbullying?" are listed in Table 2. Having consequences or punishments for offenders, monitoring and having technological systems to identify the offenders, educating students about the effects of cyberbullying, and parental involvement in teenagers' life were the strategies suggested by students to prevent cyberbullying.

Table 2: What do you think should be done to prevent cyberbullying?

Response	Percent
Consequences/Punishment	27% (204)
Unspecified Consequences/Punishment	12% (89)
Take away technology or restrict technology use	8% (63)
Jail/Juvenile Hall	2% (16)
Laws against cyberbullying	3% (25)
Fines	2% (11)
Don't know	17% (128)
Monitoring and technological identification of offenders	16% (119)
Education	12% (94)
Parental Involvement	7% (51)
No answer	6% (48)
Nothing	4% (34)
Do not use Technology/Eliminate websites	2% (11)
Other	10% (74)

The majority of the respondents believed that there needs to be a penalty to deter cyberbullying. Consequences or punishment (12%), not being allowed to use the computer or having restricted access (8%), laws against cyberbullying (3%), fines (2%), and Jail/Juvenile Hall (2%) were penalties identified by 27% of the respondents. About half of the respondents that believed that there should be a consequence or punishment, did not specify what the consequence should be. They simply stated that there should be strong punishments, penalties, rules, or consequences. Some responses that were given were: "punishment"; "there should be a serious or several serious and inescapable consequences"; "make consequences stronger"; "more strict penalties"; "stronger rules"; "stiff punishment"; and "there should be hard punishment for being caught especially if they cause the victim to commit suicide". Having mobile phones, Internet access, computer, or technology privileges revoked or suspended was the most commonly identified consequence by one third of the respondents who said there should be a consequence. Respondents said: "take their computer away"; "taking away the person's mobile phone, computer, disallowing them to visit certain websites and hoping that they grow up"; "take away their privileges"; "take away computer and mobile"; "phone use is taken away from them and loss of computer (given to police to hold onto) for awhile"; and "the teenager should be banned from using the Internet or mobile phones".

The second most common response to the question was that the respondent did not know nor was not sure what should be done. Other comments such as "if I knew I would tell you" and "society in general is going to hell, I really do not know" express a sentiment that the respondents simply did not know what to do to prevent cyberbullying. The word "nothing" was stated by 4% of the respondents. It was uncertain whether nothing means that nothing should be done to prevent cyberbullying or that nothing can be done. In some cases respondents specifically said that they felt nothing could be done.

Monitoring Internet use and having the technological means to identify offenders was the second most named strategy, identified by 14% of the respondents. Respondents said that schools, parents, and websites should be responsible for monitoring. Several respondents believed that solving the problem requires involvement of parents, teachers, websites, and law enforcement. Examples of how monitoring should be done were: "parents and schools monitoring computer use"; "screening constantly online"; "better monitoring on the Internet"; and "a monitoring system for children under 18 years old". Myspace.com and Facebook.com were mentioned as sites that should be monitored. Respondents said that there should be "monitors on website such as MySpace and Facebook that view and make sure no one is being bullied" and "I personally think there should be monitors watching all rooms, chat and MySpace". Some respondents specifically identified parents as being responsible for monitoring these activities by saying there should be "more parental controls on children's computers", "stricter guidelines from parents", and "parents should talk to their kids about it".

Along with monitoring, students suggested catching the bullies by having a technological system to trace and identify them: "Be able to trace communication so it's not anonymous, if necessary"; "easier identification of senders of Internet messages"; "a way to track all message"; "an audit trail to the person via sign on"; and "ID of the sender should be shown all the time". To deter students from cyberbullying "let the students know that everything they do on the computer can be traced."

Education was the third most mentioned strategy, supported by 12% of the respondents. Students thought there should be classes and programs at school and that cyberbullying should be "talked about at school." Hence, the schools should be responsible for providing the education. Some respondents were specific in stating that students need to know about the effects of cyberbullying. A perceptive respondent stated: "I think that people need to be better educated about this. Kids my age do stupid things because sometimes they think it's fun or they think it's just a joke. They are just kidding. They don't always see what the other person feels". Another respondent thought the education should begin in elementary school: "I think children in elementary and middle school should learn about cyberbullying, the effects, what it is, and especially the consequences. I believe if they learned what would happen to them if they are being cyberbullied, they wouldn't do it". Other respondents said: "I think kids should have to sit through a class showing the effects of what cyberbullying has done to other kids." Other solutions would be: "programmes in school"; "talk about it more at home and school"; "educate teens just how serious it is"; and "educate people to see others viewpoints".

Other respondents identified a need for students to know how victims feel: "let them see how this hurts other people"; "show the consequences"; and "educate teens just how serious it is". The students said that education also needs to include parents. They commented that "making parents aware of this, was necessary". Finally some respondents mentioned to have a "popular celebrity to campaign against cyberbullying" and "have programs come to their school and television shows teach about cyberbullying".

Parental involvement that included parental monitoring, supervision, setting consequences for cyberbullying, communication about Internet activities, and involvement in the teenager's life was reported by 7% of the respondents. Most of the respondents, who identified parental involvement, identified other strategies as well. One student quantified the need for parental involvement by saying, "Parents need to be aware of what their children are doing. If it's happening 9 out of 10 times it's going to be at their house." Another respondent described how adults should listen to teenagers:

> Adults should talk to teens better and also learn how to listen. Sometimes teens that are being bullied don't trust an adult because adults aren't as understanding as they think they are. Most adults tend to forget how it feels like to be a kid. Teens don't want to be called "snitches". Some teens may also think they can be popular by being the "funny guy" cyberbully to make others laugh. Cyberbullies should lose privilege to Internet use.

Not using technology or getting rid of websites that promote cyberbullying were preventative measures mentioned by 2% of the respondent. If no one used technology, cyberbullying would not exist, but this is not a realistic solution.

Responses from the "other category" were reporting the incident, finding help for the victim, censoring websites, don't cyberbully, get along with others, anything/something/everything, and blocking content. Each of these answers was given by less than 2% of the respondents with all of the answers totalling 10% of the respondents. The differences of opinions between bullies and victims towards a variety of punishments has been analysed in Kraft and Wang's 2009 article (Kraft & Wang, 2009) and summarized in the previous section on "Students' Perceptions of Intervention Strategies from the Research Studies".

Effects of Having Prevention Policies

In some states schools are required to have written policies that state that cyberbullying will not be tolerated. The researchers wanted to determine if teenagers that go to schools with policies against cyberbullying, were less likely to be victims or perpetrators of cyberbullying than those who go to schools that do not have the policies. The students were asked on the survey: "Does your school have a written policy not allowing online bullying?" They were given the choices of yes, no, or don't know. Based on self reported responses to questions "Have you bullied anyone online in the past year?" or "Have you been a victim of cyberbullying in the past year?" the researchers classified students as offender or victim. Offenders were students who said they bullied online. Victims were students who said they had been bullied online. Non-offenders were students that did not report bulling others online. Non-victims were students who said they were not bullied online.

Table 3 summarises students' responses to whether their school has a cyberbullying prevention policy or not versus their victim/offender status. The percentage of students who have ever cyberbullied from a school with a cyberbullying prevention policy was lower than the percentage of students from a school without a policy. Also, the percentage of students who have ever been cyberbullied, from a school with a cyberbullying prevention policy was also lower than the percentage of students from a school without a policy.

Table 3: Having a policy in school or not versus victim-offender categories

	Having a policy	Having no policy	Don't know	Total
Offenders	26.9% (94)	32.6% (114)	40.5% (142)	100% (350)
Non-offenders	23.4% (85)	26.4% (96)	50.1% (182)	99.9% (363)
Victims	26.7% (93)	31.5% (110)	41.8% (146)	100% (349)
Non-victims	23.6% (86)	27.5% (100)	48.9% (178)	100% (364)

To check whether those differences were statistically significant, the Pearson's Chi-square test method was used. The null hypothesis was: The percentage of offenders and victims attending schools with a cyberbullying policy are the same as the percentage of offenders and victims, respectively, attending schools without a cyberbullying policy. The derived $\chi 2$ value was 6.136 with degree of freedom 1, so the p-value was 0.02. Therefore, the null hypothesis was rejected.

The result of the Chi-square test shows that the percentage of offenders in schools with cyberbullying prevention policies is statistically lower than the percentage of offenders in schools without the policies. The percentage of cyberbullying victims in schools with cyberbullying prevention policies is also statistically lower than the percent of victims in schools without the policies. Therefore, according to this study, cyberbullying prevention policies reduce cyberbullying activities as well as the possibility that teens are harassed by cyberbullying. This finding is consistent with past research findings that the key component to a successful bullying programme is to have a policy that clearly states a school's position against bullying (Smith *et al.*, 2004).

Who Is Responsible for Preventing and Stopping Cyberbullying?

The students were asked: "Whose responsibility do you think it is to have rules and penalties for teens to stop them from cyberbullying?" Students could select more than one choice from the seven stakeholders. Table 4 summarises students' answers to this question in terms of the percentage of the total number of responses and in terms of the percentage of 713 students that thought each stakeholder was responsible.

Table 4: Whose responsibility is it to stop cyberbullying?

Who takes responsibility?	Total	Percent of Total Number of Response	Percent of 713 Respondents
Parents	612	23%	86%
Schools	457	17%	64%
Social Networking websites	385	15%	54%
Laws	363	14%	51%
Teens tell adults about bullying incidents	319	12%	45%
Any type of websites	276	11%	39%
Internet service providers	213	8%	30%

The seven choices of stakeholders in Table 4 are ranked according to the number of students who checked yes, as shown in the total column on the right. The percentage of the total number of responses for each stakeholder is listed in the column next to the stakeholder. Since students could identify more than one stakeholder the percent of the 713 students that identified each stakeholder is listed in the final column. Students thought parents were the most responsible stakeholder for stopping cyberbullying, followed by schools, social websites, and law, all of whom were voted for by more than 50% of 713 students in the survey.

The respondents ranked parents as highest and schools as second highest stakeholder responsible for having rules and penalties for teenagers to stop them from cyberbullying. This finding is consistent with findings that parental involvement and supervision is a key component of preventing cyberbullying (Tully, 2007; Willard, 2007a). Having schools contact parents to informally resolve a cyberbullying incident

(Tully, 2007; Willard, 2007b) gives parents the opportunity to be the primary stakeholder followed by schools to stop and prevent cyberbullying which is how the respondents in this study perceived the hierarchy of responsibility.

How Do Students Want to Report a Cyberbullying Incident?

In the survey students were asked "How do you want to report a cyberbullying incident?" They chose an answer from five choices. The results are listed in Table 5.

Table 5: How do students want to report a cyberbullying incident?

	Total
Tell your parents	41% (292)
Send an e-mail to a central e-mail address that handles cyberbullying incidents	27% (194)
Call an anonymous hotline that handles cyberbullying incidents	17% (122)
Tell an adult at school	9% (67)
Other	5% (37)

The results show that students would like to report cyberbullying incidents to their parents followed by sending an e-mail to a central e-mail address that handles cyberbullying incidents, calling an anonymous hotline, and telling an adult at school. The finding is similar to the results of a study by Agatston (2007) where students reported that they would be more likely to report a threatening incident of cyberbullying to their parents than to an adult at school because they do not think the adults at school could help them with the cyberbullying problem (Agatston, 2007). Sending an e-mail was the second most favoured method. A respondent in the study expressed a need for a place to report incidents as the person believed that having a reporting system could help with prevention. The respondent said, "I think that an open place to report cyberbullying could help. If a victim feels that by speaking up it could stop and be taken seriously, there would be more people speaking about bullying."

Table 6 shows the differences of the preferences to who a cyberbullying incident is to be reported between bullies and victims. Students are grouped into bullies and non-bullies. Bullies include students who had experiences with bullying others. A bully can be a pure offender or both a bully and a victim. Non-bullies include students who did not have experience with bullying others. A non-bully can be a victim or neither a bully nor a victim. The analyses in Table 6 show that more non-bullies (44.9%) versus 37.14% of the bullies prefer to report to parents; while more bullies (29.14%) versus 25.34% of the non-bullies preferred to report to a central place online that handles cyberbullying. The difference of percentages between bullies and non-bullies on preference of telling parents, namely 7.76% is statistically significant with a p-value of 0.0174. The other differences are not statistically significant.

Table 6: Perspectives of bullies and non-bullies on whom to report cyberbullying

Whom to report	Bullies		Non-bullies		% difference significant?
Tell parents	130	37.14%	163	44.90%	Yes (p=0.0174)
Send e-mail to an address handling cyberbullying	102	29.14%	92	25.34%	No (p=0.1271)
Call an anonymous hotline handling cyberbullying	58	16.57%	63	17.36%	No
Tell an adult at school	33	9.43%	34	9.37%	No
Other	27	7.71%	11	3.03%	No
Total	350	100%	363	100%	

Discussion

Seventy-two percent of the 713 respondents viewed cyberbullying as being wrong and they want schools or parents to do something, mild or severe, to stop cyberbullying. Almost half of the respondents thought that teens should be punished after being caught cyberbullying for the first time. Two-thirds of the respondents who recommended specific punishments suggested that the punishment would be to revoke the offender's technology privileges. Only five percent of the students surveyed, thought nothing should be done about cyberbullying. When schools are setting up cyberbullying prevention policies, the data in this survey is telling them a strong message that students expect punishments for cyberbullying, even for first-time offenders. However, schools tend to be reluctant to punish students for cyberbullying because they are afraid of violating their free speech rights. Only one student in the survey mentioned Free speech rights as an issue. If a school was thinking solely about students' acceptance of a policy on preventing cyberbullying, then our survey results would say that the majority of the students' would be in favour of school enforcing consequences for cyberbullying. However, in reality, when students get into trouble for cyberbullying their parents often seek legal counsel to settle the issue. The school is then in jeopardy of losing the case based on Free Speech rights of the offender.

Schools are allowed to discipline student for off campus speech that meets the Tinker test of *material disruption* (Tinker v. Des Moines,

1969). The Tinker case is over 40 years old and does not reflect the role of the Internet in today's society. It is argued that laws will not solve the cyberbullying problem because they do not teach students why they should not cyberbully (Szoka & Thierer, 2009). In order for students to learn why they should not cyberbully there need to be enforceable consequences when they are caught cyberbullying. With the current legal climate parents are responsible for monitoring their teenager's online activities and enforcing consequences for cyberbullying behaviour. The consequence teenagers favour the most is revoking technology privileges. The parents do not understand the impact of taking away a teen's technology privileges. They think their teenager needs the Internet for school work when, in fact, most Internet use at home is for socialising (Brown, Jackson, & Cassidy, 2006). By taking away access to Internet and technology teenagers lose their primary means of communication in their social life (Brown *et al.*, 2006). It makes sense that this consequence could be a deterrent to cyberbullying.

Schools are currently trying to prevent cyberbullying by having policies against it, educating students about the effect of cyberbullying and having Internet safety programs. But Internet safety programs have not been effective in stopping and preventing cyberbullying. The content of the education needs to change in order for students to understand the impact of cyberbullying on the victim. The students in this survey specifically addressed the importance of letting teenagers know how serious cyberbullying is. Since offenders cannot see the reaction of their victims, it is difficult for them to assess the impact of their actions.

According to the results of our study about half of the non-offenders and non-victims did not know if their school had a policy against cyberbullying. More offenders and victims knew about school policies, probably because they have been impacted by them. Our results showed that cyberbullying prevention policies do reduce bullying activities as well as decrease the possibility that teenagers are harassed with cyberbullying. Schools are on the right track by establishing cyberbullying prevention policies. But they do need to improve the communication of these policies to students.

Limitations of This Study

A major limitation of the study is that the researchers never really knew who was doing the survey. Also, subjects may not be completely honest in their responses. Online surveys are convenience surveys. They are not necessarily representative of a larger population of teenagers or online teenagers. However the study has specific value as it provides insight in the way how teenagers view cyberbullying prevention strategies.

In this study the investigators classified offenders and victims of cyberbullying as respondents who reported at least one incident of cyberbullying. We did not consider reports of the severity and frequency of the incidents. We did ask respondents how often they were cyberbullied as other researchers (Li, 2007) found that most subjects being victims or offenders of cyberbullying reported it only once or twice.

The responses to the open ended questions were coded by the researchers. The coding is subject to interpretation by the researchers. Not every subject provided a solution to answer this question. "I don't know" was the response for 17% of the classification questions and N/A for 6%.

The market research firm estimates that there is an error rate of 5-6% for this survey. Technological measures were taken by the market research firm to eliminate common markers of erroneous data. As these measures cannot be taken by investigators who use paper surveys, it gives value to having the survey administered online by a market research firm as opposed to paper surveys.

Conclusion

Successful prevention of cyberbullying involves cooperation among teenagers, their parents, schools, law enforcement and website providers. Based on the data collection from a nationwide sample of United States teenagers between 13 and 17 years old, we learned that 72% of teenagers believe that something needs to be done when a teenager is caught cyberbullying for the first time. The results showed with statistical significance that cyberbullying prevention policies do have the effects of reducing bullying activities as well as reducing the possibility that teenagers are being harassed by cyberbullies. Students thought that parents are the primarily stakeholders, followed by schools, responsible for preventing cyberbullying. It is therefore recommended that schools develop policies that set clear rules and consequences for cyberbullying. These policies should be communicated to students and parents. Schools should consider having an anonymous central e-mail where students can report cyberbullying incidents, as less than 10% of students selected "telling an adult at school" as a way they would report cyberbullying incidents. It is recommended that parents become more educated about Internet safety and monitor their children's Internet usage. Parents should openly discuss ways of handling cyberbullying incidents with their children as they are considered to be primarily responsible for preventing and stopping cyberbullying of their teenagers.

References

Agatston, P., Kowalski, R., & Limber, S. (2007). Students Perspectives on Cyberbullying. *Journal of Adolescent Health, 41*(6), 59-60.

Aftab, P. (2010a). *Stop Cyberbullying; What's the Parent's Role in This.* Retrieved December 17, 2009, from http://www.stopcyberbullying.org/ index2.html.

Aftab, P. (2010b). *What Is Cyberbullying Exactly?* Retrieved April 23, 2010, from http://www.stopcyberbullying.org/index2.html.

Associated Press. (2005, April 5). *N.J. School Rapped for Punishing Student over Web Site.* Wire report. Retrieved June 10, 2007, from http://www.firstamendmentcenter.org/n-j-school-rapped-for-punishing-student-over-web-site.

Ballou, B. & Ellement, J. (2010, March 29). Nine Charged in Death of South Hadley Teen Who Took Life After Bullying. *Boston Globe*. Retrieved April 25, 2010, from http://www.boston.com/news/local/breaking_news/2010/ 03/holding_for_pho.html.

Belsley, B. (2005). Cyberbullying: An Emerging Threat to the "Always ON" Generation. *Technology Concerns, 1*(2), 6-13.

Belsey, B. (2006). Bullying.org: A Learning Journey. *Bulletin – Newfoundland and Labrador Teachers Association, 49*(4), 20.

Belsey, B. (2010). *Www.cyberbullying.org: Always On, Always Aware*. Retrieved April 30, 2010, from http://www.cyberbullying.org/.

Beran, T. & Li, Q. (2005). Cyberharassment: A Study of a New Method for an Old Behaviour. *Journal of Educational Computing Research, 32*(3), 265-277.

Berson, I.R., Berson, M.J., & Ferron, J.M. (2002). Emerging Risk of Violence in the Digital Age: Lessons for Educators from an Online Study of Adolescent Girls in the United States. *Journal of School Violence, 1*(2), 51-71.

Brown, K., Jackson, M., & Cassidy, W. (2006). Cyber-bullying: Developing Policy to Responses That Are Equitable and Effective in Addressing This Special Form of Bullying. *Canadian Journal of Educational Administration and Policy, 57*, 8-11.

Bullypolice.org. (2010). *Forty-three States Have Anti-bullying Laws*. Retrieved June 17, 2010, from http://www.bullypolice.org/.

Campbell, M. (2005). Cyberbullying: An Older Problem in a New Guise? *Australian Journal of Guidance and Counselling, 15*(1), 68-76.

Chibnall, S., Wallace, M., Leicht, C., & Lunghofer, L. (2006). *I-safe Evaluation.* Final Report. Rockville, MD: National Institute of Justice/NCJRS.

Child Health Research Promotion Center. (2009). *Australian Covert Bullying Prevalence Survey.* Retrieved November 24, 2010, from http://www.deewr. gov.au/Schooling/NationalSafeSchools/Pages/research.aspx.

Collerton, S. (2009, July 23). Get Tech Savvy to Stop Cyber Bullies, Parents Told. ABC News. Retrieved July 24, 2009, from http://www.abc.net.au/ news/stories/2009/07/23/2634545.htm.

Conklin, M. (2009 April). *What Impact Do "Bad Respondents" Have on Business Decisions? A Market Tools White Paper*. Retrieved June 21, 2010, from http://www.markettools.com/KnowledgeCenter.

Cowie, H. & Colliety, P. (2009). *Cyberbullying – The Situation in the UK- Country Report.* University of Surrey, United Kingdom. Retrieved November 24, 2010, from http://www.cybertraining-project.org/page.php?lang=En&page=8.

Crimaldi, L. (2010, March 29). DA: School Knew of Brutal Bullying of Phoebe Prince. *Boston Herald.* Retrieved April 24, 2010, from http://news.bostonherald.com/news/regional/view.bg?articleid=1243175&format=comments#CommentsArea.

Crombi, G. & Trinneer, A. (2003). *Children and Internet Safety: An Evaluation of the Missing Program.* A Report to the Research and Evaluation Section of the National Crime Prevention Centre Justice of Canada. Ottawa: University of Ottawa.

Feinberg, T. & Robery, N. (2009). Cyberbullying: Intervention and Prevention Strategies. *Communique*, *38*(4), 1-5.

Franek, M. (2005/2006), Foiling Cyberbullies in the New Wild West. *Educational Leadership*, *63*(4), 39-43.

Halligan, J. (2010) *Ryan's Story Presentation.* Retrieved April 28, 2010, from http://www.ryanpatrickhalligan.com/.

Internet Safety Technical Task Force. (2008). *Enhancing Child Safety and Online Technologies.* Final report of the Internet Safety Technical Task Force to the multi-state working group on social networking of state attorneys general of the United States. Retrieved April 30, 2010, from http://cyber.law.harvard.edu/pubrelease/isttf/.

Juvonen, J. & Gross, E. (2008). Extending the School Grounds? Bullying Experiences in Cyberspace. *Journal of School Health*, *78*(9), 496-505.

Keith, S. & Martin, M. (2005). Cyber-Bullying: Creating a Culture of Respect in a Cyber World. *Reclaiming Children and Youth*, *13*(4), 224-228.

Kowalski, R. (2008) Cyber Bullying: Recognizing and Treating Victim and Aggressor. *Psychiatric Times*, *25*(11).

Kraft, E. (2006). Cyberbullying: A Worldwide Trend of Misusing Technology to Harass Others. In K. Morgan, C.A. Brebbia, & K.M. Spector (eds.). *WWW the Internet Society II: Advances in Education, Commerce & Governance* (pp. 155-166). Southampton, England: WIT Press.

Kraft, E.M. & Wang, J. (2009). Effectiveness of Cyberbullying Prevention Strategies: A Study of Student's Perspectives. *International Journal of Cybercriminology*, *3*(2), 513-535.

Layshock v. Hermitage School District 2:06-cv-116, United States District Court for the Western District of Pennsylvania, 2006 U.S. Dist. LEXIS 21080, April 7, 2006, Decided, Reconsideration denied by Motion denied by Layshock v. Hermitage Sch. Dist. 2006 U.S. Dist. LEXIS 20354 (W.D. Pa., Apr. 13, 2006).

Li, Q. (2007). New Bottle but Old Wine: A Research of Cyberbullying in Schools. *Computers in Human Behaviour*, *24*(4), 1771-1791.

Li, Q. (2010). Cyberbullying in High Schools: A Study of Students' Behaviours and Beliefs About This New Phenomenon. *Journal of Aggression, Maltreatment and Trauma*, *19*(4), 372-392.

Lines, E. (2007). *Cyberbullying: Our Kids New Reality. A Kids Hulp Phone Research Study of Kids Online.* Kids Help Phone. Retrieved November 28, 2010, from http://org.kidshelpphone.ca/media/21704/2007_cyber_bullying_report.pdf.

Livingstone, S. (2006). *UK Children Go Online. End of Award Report.* Media@LSE: London. Retrieved July 26, 2009, from http://www.lse.ac.uk/collections/children-go-online/.

Livingstone, S. & Haddon, L. (2008). Risky Experiences for Children Online: Charting European Research on Children on the Internet. *Children & Society*, *22*(4), 314-323.

Livingstone, S., Haddon, L., Görzig, A., & Olafsson, K. (2009). *Risks and Safety on the Internet: The Perspective of European Children, Initial Findings from the EU Kids Online Survey of 9-16 Year Olds and Their Parents.* LSE, London: EU Kids Online. Retrieved November 24, 2010, from http://www2.lse.ac.uk/media@lse/research/EUKidsOnline/Home.aspx.

Market Tools. (2010). *Sample from ZoomPanel.* Retrieved June 21, 2010, from www.markettools.com/sites/default/files/resources/data_sheet/DS_Zoom Panel.pdf.

Market Tools, Inc. & Zoomerang.com. (2010). *Survey Respondents: Profile Reference Book.* Retrieved June 20, 2010, from www.zoomerang.com/resources/Panel_Profile_Book.pdf.

McCarthy, C. (2009, July 2). *Report: Guilty Verdict Overturned in MySpace Suicide Case*. Retrieved April 23, 2010, from news.cnet.com/8301-13577_3-10278483-36.html.

McCormick, S. (2005, November 24). Bullying: Sticks, Stones, and Web Chat. *The Independent Online Edition.* Retrieved June 22, 2010, from http://www.independent.co.uk/news/education/education-news/bullying-sticks-stones-and-web-chat-516631.html.

Media Awareness Network. (2001). *Young Canadians in a Wired World: Key Findings: Analysis of Risky Behaviors*. Retrieved November 24, 2010, from http://www.media-awareness.ca/english/research/YCWW/index.cfm.

Megan Meier Foundation. (2010). Retrieved April 23, 2010, from http://www.meganmeierfoundation.org.

Megan Meier Cyberbullying Prevention Act. (H.R. 1966, 111th Congress, 2009-2010). Retrieved May 16, 2010, from http://thomas.loc.gov/cgi-bin/query/z?c111:H.R.1966.

Mishna, F., Saini, M., & Solomon, S. (2009). Ongoing and Online: Children and Youth's Perceptions of Cyberbullying. *Children and Youth Services Review*, *31*(12), 1222-1228.

Moessner, C. (2007). Cyberbullying. *Trends and Tunes, 6*(4). Retrieved June 11, 2009, from http://www.harrisinteractive.com/news/newsletters/k12news/HI_TrendsTudes_2007_v06_i04.pdf.

Mustacchi, J. (2009). Who's Keeping Student's Safe Online? *District Administration, 45*(1), 12.

National Children's Home. (2005). *Putting U in the Picture. Mobile Bullying Survey 2005*. Retrieved November 24, 2010, from, http://www.filemaker.co.uk/educationcentre/downloads/articles/Mobile_bullying_report.pdf.

Opinion Research Corporation (2006). *Cyber Bully Teen.* Prepared For Fight Crime: Invest in Kids, July 6, 2006, Opinion Research Corporation: Princeton, New Jersey. Retrieved October 1, 2006, from http://www.fightcrime.org/cyberbullying/cyberbullyingpreteen.pdf.

Patchin, J.W. & Hinduja, S. (2006). Bullies Move Beyond the Schoolyard: A Preliminary Look at Cyberbullying. *Youth Violence and Juvenile Justice, 4*(2), 148-169.

Patchin, J.W. & Hinduja, S. (2010). Bullying, Cyberbullying, and Suicide. *Archives of Suicide Research, 14*(3), 206-221.

Protecting Children of the 21st Century Act. (S.1492, 110th Congress, 2008). Retrieved April 30, 2010, from http://www.theorator.com/bills110/text/s49.html.

Sauders, R. (2005). Happy Slapping: Transatlantic Contagion or Home-grown, Mass-mediated Nihilism? *Static Issue 1: Play and Violence*, The London Consortium. Retrieved April 30, 2010, from http://static.londonconsortium.com/issue01/index.html.

Shariff, S. (2005). Cyber-Dilemmas in the New Millennium. *McGill Journal of Education, 40*(3), 467-487.

Shariff, S. (2008). *Cyber-Bullying: Issues and Solutions for the School, the Classroom and the Home.* Canada: Routledge.

Smith, P., Peppler, D., & Rigby, K. (2004). *Bullying in Schools: How Successful Can Interventions Be?* United Kingdom: Cambridge University Press.

Stacey, E. (2009). Research into Cyberbullying: Student Perspectives on Cybersafe Learning Environments. *Informatics in Education, 8* (1), 115-129.

Szoka, B. & Thierer, A. (2009). Cyberbullying Legislation: Why Education Iis Preferable to Regulation. *Progress on Point, 16*(12), The Progress and Freedom Foundation. Retrieved April 29, 2010, from http://www.pff.org/issues-pubs/pops/2009/pop16.12-cyberbullying-education-better-than-regulation.pdf.

Text Someone.com. (2010). *Text Someone Homepage*. Retrieved November 2, 2010, from http://www.textsomeone.com.

Tinker v Des Moines Independent County School District 393 U.S. 503 (1969).

Thomas, K. (2009). *Teen Online and Wireless Safety Survey: Cyberbullying, Sexting, and Parental Controls*, Cox Communications Teen Online and Wireless Safety Survey in Partnership and the National Center for Missing and Exploited Children and John Walsh. Retrieved July 23, 2009, from http://www.cox.com/takecharge/safe_teens_2009/research.html.

Tully, J.D. (2007). *The Outer Limits: Disciplining Students Without Getting Sued*, West Regional Equity Network. Retrieved June 6, 2007, from http://miketullylaw.com/library/The_Outer_Limits_Disciplining_Cyber-Mischief_Without_Getting_Sued.pdf.

Vandebosch, H. & Van Cleemput, K. (2008). Defining Cyberbullying: A Qualitative Research into the Perceptions of Youngsters. *Cyber Psychology & Behaviour*, *11*(4), 499-503.

Willard, N. (2006). Flame Retardant. *School Library Journal*, *52*(4), 55-56.

Willard, N. (2007a). *Cyberbullying and Cyberthreats: Responding to the Challenge of Online Social Aggression, Threats, and Distress.* Champaign, IL: Research Press. Retrieved July 24, 2009, from http://www.cyberbully.org/cyberbully/cbbook.php.

Willard, N. (2007 b). Cyberbullying Legislation and School Policies: Where Are the Boundaries of the "Schoolhouse Gate" in the New Virtual World?, Paper, Center for Safe and Responsible Use of the Internet. Retrieved April 29, 2010, from http://www.cyberbully.org/cyberbully/docs/cblegislation.pdf.

Willard, N. (2007c). Educator's Guide to Cyberbullying and Cyberthreats. Center for Safe and Responsible Use of the Internet. Paper, Center for Safe and Responsible Use of the Internet. Retrieved December 16, 2009, from http://cyberbully.org/cyberbully/docs/cbcteducator.pdf.

Ybarra, M. & Mitchell, K. (2007a). Prevalence and Frequency of Internet Harassment Investigation: Implications for Adolescent Health. *Journal of Adolescent Health*, *41*(2), 189-195.

Ybarra, M., Diener-West, M., & Leaf, P.J. (2007b). Examining the Overlap in Internet Harassment and School Bullying: Implications for School Intervention. *Journal of Adolescent Health*, *41*(6), 42-50.

ZoomPanel.com. (2010). Retrieved June 21, 2010, from http://join.zoompanel.com/zoompanel/index.php?menu=About+Us.

Fotologs and Love Socialisation Processes

A Conventional or a Transformative Model of Sexuality and Relationships?

Iolanda TORTAJADA, Cilia WILLEM,
Lucrezia CRESCENZI, Núria ARAÜNA & Itxaso TELLADO

The Role of Media in Gender Violence

This paper aims at gaining insight into the link between media uses and causes of gender violence and, specifically, at assessing whether social networking sites (hereafter SNS) can promote this kind of violence or rather can contribute to eradicating it. In this sense, we are interested in the role of representations and media practices in love socialisation processes. The study draws upon two fundamental ideas: on the one hand, the idea that some teenagers link attraction to violence, and that such a link promotes relationships in which violence may emerge (Valls, Puigvert, & Duque, 2008), and on the other hand, the notion that sexual and affective relationships are based on attraction and processes of choice in which the kind of values that we internalise are informed by referents from the media (Gómez, 2004). Where do our ideas about love and attraction come from? How are they constructed and sustained? What do the media have to do with it?

The reflections and results that are presented here are framed in the research project "Media and gender violence: catalysts or factors of prevention?', funded by the Spanish Ministry of Science and Innovation (2009-2011).

The Socialisation of Love as a Starting-Point

According to Oliver and Valls (2004), falling in love is not as spontaneous as we tend to think. Rather, we feel attracted by individuals whom we believe are valuable as lovers and transform them into something desirable. We therefore need to ask ourselves if affection is inevitable and if attraction is something that happens without a reason, given that ethics often provides us with egalitarian values for relationships, whereas desire can make them extremely unequal and damaging in practice. In this paper we focus on what Gómez (2004) calls *love socialisation processes*.

In love socialisation processes, multiple socialisation factors are involved, such as the family, formal and informal educational institutions, the media and, – very strongly during adolescence, – peer groups: "[…] attraction, desire and love are not automatically given, but they are the result of social interactions that are reconstructed by each one of the participants" (Oliver & Valls, 2004, p. 106; our translation from original quote).

But what kind of attraction and love models do we find in peer interactions between youngsters? Although interpersonal and intergroup interaction is variable when considering ways of understanding sexual and affective relationships, teenagers generally still seem to respond to the most traditional patriarchal schemes when asked about their models of attraction (Padrós, 2007). The discourse of the boys quite explicitly refers to physical aspects when assessing female attractiveness (beauty, or topics related to sex): boys still go for *showpieces* or consider women as objects, or want the *unavailable woman* (by unavailable they understand the one desired by everybody, or the girl who is in a higher class at school, which makes it unrealistic to expect her to even look at them). There is also a general consensus that boys who treat their girlfriends badly or cheat on them are perceived as the most attractive by girls: they indicate that they would be *turned on* by those boys, and most of the boys assume that if they want to be popular they have to behave like that. Thus, the *naughty boy* is a fashionable social category, while the "good boy", who has a caring attitude and listens to women, is dissociated from attraction or any potential love affair and classed in the category *just a friend* (Padrós, 2007). Moreover, attraction is linked – both for girls and boys – to specific dress and hairstyles that are "cutting-edge" in their own terms.

However, it has to be said that, apart from this traditional model of attraction, there are ways of relating to each other, based on respect and equality, that build an alternative model of sexual and affective relationships (Gómez, 2004; Duque, 2006).

Traditional and Alternative Models of Attraction and Choice

Departing from the complexity and diversity of affective and sexual relationships shown throughout history until today, Gómez (2004) tries to find some common traits in the understanding of the role models for love and sex that may promote violence. He comes up with the traditional model of sexual-affective attraction and choice, an abstract model that, according to the author, is a result of patriarchal and hierarchical values that stem from discriminatory and individualistic societies. With respect to sexual and affective relationships, this model links desire to violent attitudes, to contempt and to a lack of commitment, thus dissociating desire and affection and encouraging attraction to power, when assessing male attractiveness, and to beauty, when assessing female attractiveness. Furthermore, attraction is focused on the position of "hard to get", and on the character of the "womaniser". Some women imitate these masculine models, not allowing egalitarian relationships and increasing suffering for those involved in traditional models (Giddens, 1995). Therefore, care and respect are devoid of potential excitement and this induced thought is perceived as natural: "we are encouraged (and thus bound) to look for someone who is out of reach, violence-prone and yet highly desirable" (Gómez, 2004, p. 73). In this model, love is considered as "inexplicable" and "out of control", based on instinct and biology and therefore unavoidable. Following the ideals of romantic love, it is supposed to last forever and to be "unique" (Giddens, 1992). Some of these ideas have equally been pointed out by Mary Lou Galician (2004) when she analyses the (harmful) myths of love perpetuated by the media. Gómez also includes in this model the stable and boring couples who assume that if they want a long-term and quiet relationship they will have to give up passion. According to Gómez, de Botton and Oliver (2009), this model is sustained by the unequal distribution of roles between men and women in traditional societies.

Gómez (2004) goes on to propose an alternative model of sexual-affective attraction and choice, one which links passion to tenderness and care. This model, later referred to as *radical love* (de Botton & Oliver, 2009), implies that sexual and affective relationships are established in a communicative negotiation on equal terms. This model is based on the idea of a reflexive and democratic society that has maximised choices for all its members. The reflexivity it implies (Beck & Beck-Gernsheim, 1998) is typical of modernity and allows subjects to redefine intimacy and its rules. In this process, inter-subjectivity is the key to the construction of meaning and to making sense of life (Gómez, 2004). Romantic ideals can be sustained in this model as long as they satisfy both partners (de Botton & Oliver, 2009). In any case, passion is not considered as incompatible with stability. This idea of a progressive love model corresponds to the depictions of Giddens (1992) when he describes *confluent love*: active and contingent rather than ethereal. This love then leads to a "pure relationship" in which both subjects establish a non-utilitarian link based on dialogue and mutual support.

Photoblogs as Mediators of Intimacy

Web 2.0 and Self-disclosure

The Internet, with its 2.0 tools, has become a privileged instance of self-reflection and peer interaction among teenagers: "teens are modelling identity through social network profiles so that they can write themselves and their community into being" (Boyd, 2007, p. 2). SNS are interesting for our study because, as Boyd and Ellison (2007) have noticed, they are generally changing from a model based on common interests to one based on people and relationships. This means that these sites are organised as *egocentric* networks, offering users a communication context where we as researchers can find portrayals of sexual and affective relationships made by youngsters. SNS promote what has been called *intimate enhanced self disclosure* (Valkenburg & Peter, 2009), as they shift the social space of interactions from the public sphere to the more reduced and controlled universe of the private bedroom. In these self-generated media teenagers easily share their intimate experiences and learnt ways of portraying love and gender roles, and cyberspace becomes a site of meanings about attractiveness and sexual and affective relationships. Because of the content that is distributed across this medium, it acts as a *super peer* in the way it was defined by L'Engle, Brown and Kenneavy (2006) after studying teenagers in search for information about sex on the Internet. The authors concluded that this platform provided the information required by users while allowing for anonymity.

Thus the relevance of SNS in socialisation, and specifically in gender identity construction, has already been largely analysed (Ringrose, 2010; Sevick Bortree, 2005; Sveningsson, 2007; Huffaker & Calvert, 2005; Thelwall, 2008). The role of SNS in love and sex socialisation processes has not yet attracted a significant body of research. A relevant departure point is to consider SNS' in the same way that Gill (2009) frames other media: as *mediators of intimacy*, meaning that these media play a role in promoting specific behaviours and beliefs concerning sexual and affective relationships.

In order to deal with these issues we have chosen Fotolog, a virtual platform that displays not only texts but also images created by its users. On this platform, as on other 2.0 sites, youngsters are not just consumers but active *prosumers*, as they create their own content (McLuhan & Nevitt, 1972; Tapscott & Williams, 1996) or, as recently theorised, *produsers* (Bruns, 2005). As mentioned above, the material they *produse* is mainly related to their own affective and sexual experiences, which makes it fit for analysis in the context of our research questions. The images youngsters upload are also epistemologically interesting, as teenagers try to portray themselves in ways that make them attractive to others (mainly their peers) and desirable for relationships (Sevick Bortree, 2005).

The Hyper-ritualisation

We study these self-representations following the concept of *hyper-ritualisation* as described by Goffman (1991). According to this author, there are plenty of ritualisations and liturgical situations in social behaviour aimed at facilitating mutual orientation of participants in social interaction. For Goffman (1991), an example of hyper-ritualisation is when advertisers use poses and attributes in their advertisements in order to display extreme normalisation, exaggeration and simplification: "If anything, advertisers conventionalise our conventions, stylise what is already a stylization, make frivolous use of what is already something considerably cut off from contextual controls. Their hype is hyper-ritualisation" (Goffman, 1979, p. 84). Applied to our research, this idea of hyper-ritualisation can show how teenagers take ritual referents from situational definitions already established (partly by their peer group) and use them to *hyper-ritualise* themselves and their concerns in a series of predictable images. For example, there seem to be very concrete patterns of representation as to what is supposed to be *sexy* among certain groups of teenagers. In Fotolog, the gestures, clothing, putting on faces, adopting poses and showing certain parts of the body seem to be previously codified categories, even if in the end every subject decides the *mise en scène* and chooses the photos she or he will use for self-representation. Depictions that are considered appropriate and meaningful by users are those based on profiles and self-portrayals of other (Boyd, 2007). The recurrence on the displays of different users has been defined as a kind of *template* (Ringrose, 2010) that makes every media autonomous even if it takes codes of genres such as advertisements or celebrity culture.

Furthermore, Fotolog is an SNS that allows for sharing meanings, so we can assume a certain degree of consensus regarding gender relationships, or at least a tendency to that consensus, which would be built through the inter-subjective processes that take place in Fotolog. We should also draw attention to the conventionalisation of representational modes in this platform: the textual and visual styles of users are so similar that it raises the question whether the same youngsters would portray themselves differently in another context. In this sense it can be argued that specific genre codes are applied to a specific platform (in this case Fotolog), but in any case it allows us to observe some traits referring to gender identity and concepts of love and sexual relationships. In other words, it seems to make sense to say that Fotolog creates a series of formats that respond to a specific *media logic* (Altheide & Snow, 1979) and to the expectations of the peer group in terms of attractiveness.

SNS constitute a complex field of meanings related to gender portrayals and intimate relationships based on a combination of existing peer-culture and the technical possibilities inherent in the medium (Livingstone, 2009).

Research Questions

This analysis focuses on how youngsters represent their relationships and desires. This study looks at whether teenagers use this new medium to reproduce traditional models of relationships and gender portrayals or if, on the contrary, the broadens their opportunities to build a transformative model encouraging gender equality and the democratisation of personal life (Gómez, 2004; Giddens, 1995).

The following research questions summarise the study's aims:

RQ1: What kind of self-representations do boys and girls choose in their Fotologs when they intend to be attractive to others?

RQ2: What kinds of ritualisation underlie the self-representations constructed by teenagers and shared in their Fotologs?

RQ3: Which love socialisation model (traditional or alternative) can be revealed from teenagers' practices and interactions on Fotolog?

Method

Sample

We selected a non-probabilistic sample of 400 entries from 400 different Fotolog[1] accounts registered by teenagers that are used for self-representation and images of daily life. We excluded Fotolog accounts sharing artistic photography or that specialised in certain topics such as pets, travelling or movies. From this sample, 355 posts were effectively analysed, as 45 accounts had been closed by the end of the data-gathering period. The sample was taken from December 2009 to February 2010 and the material was analysed from March to April 2010. The fact that in less than five months 45 individuals "disappeared", points out to the ephemeral character of our object of study.

Procedures

We established four areas of analysis per case: the *usernames* teenagers choose when they sign up for Fotolog, the *pictures* they upload, the *texts* that accompany those pictures, and the *comments* they receive from friends to their update.

1 According to the service provider, Google-owned Fotolog is the second most used SNS in Spain, and 35% of its users are aged 15 to 24.

Image 1: Areas of analysis: username, picture, text and comments of Fotolog entries

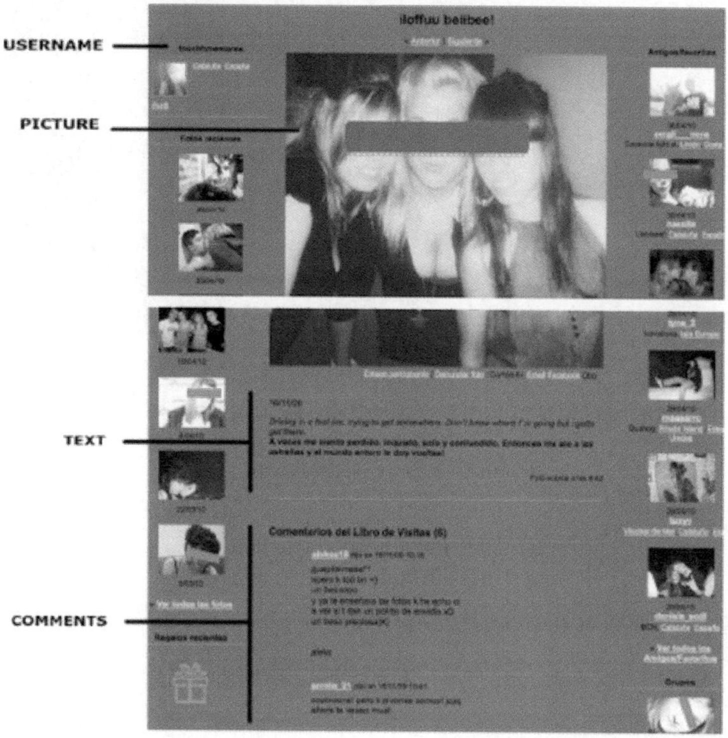

Taken from fotolog database (F20, female)[2]

An online analysis tool was created specifically for the research,[3] which allowed the team to work on the cases simultaneously.

---

[2] Every fotolog entry has been coded with a capital "F" (for "fotolog") followed by a number corresponding with its (arbitrary) place of order in our sample. The gender of the fotolog account holder is also added. In a few cases, texts quoted belong to commentators to an account holder's entry.

[3] The authors would like to thank Antonio Bartolomé at the Interactive Media Lab for developing the analysis tool.

Image 2: Example of a record in the online analysis tool

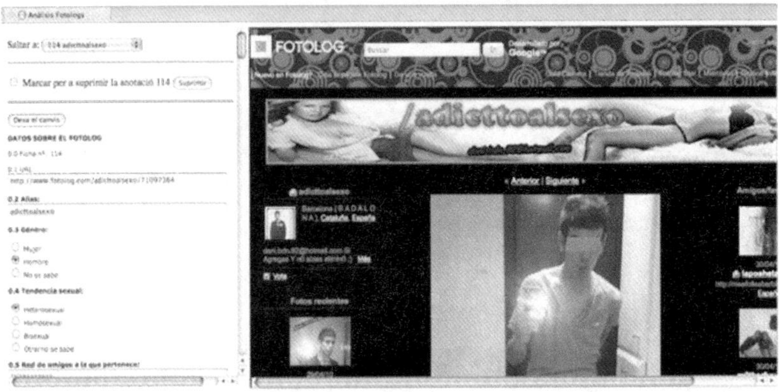

We applied a critical discourse analysis to the active sample, mainly (although not exclusively) informed by previous theoretical categories. Each unique Fotolog entry was analysed quantitatively, but when significant data emerged, other entries of the same user or responses from other users to that particular post were analysed in a more qualitative and critical way.

Results: Hyper-ritualisation in Practice

Youngsters use Fotolog intensively: most accounts have daily entries, and 95% of users have uploaded pictures in the last two months. The entries or posts are also referred to in this paper as *updates*. Only one picture can be uploaded per update. 15% of the girls and 67% of the boys have 50 or more updates since their first login on Fotolog. Each profile receives an average of 17 comments to their updates (boys receive an average of 20 and girls an average of 16). In some cases updates received more than 200 comments.

Usernames

Among the 400 profiles that we analysed, 25% of usernames had a direct relationship with the user's (presumed) real name, for example, *nuuuuuriAa* for "Núria". The "distortion" in the spelling of names may be due to the fact that Fotolog is a public SNS and teenagers want to avoid being identified by adults. Surprisingly, 20% of usernames had explicit sexual connotations. The rest were combinations of names of friends or couples, declarations of love (*you_are_my_life*, *i_need_your_love*, or resentment (*get_out_of_my_life*, *you_are_all_the_same*).

Unsurprisingly, we were most interested for this study in usernames that implied sexual connotations. When teenagers "rename" themselves, they use expressions like: *what_if_i_violated_you* (male) or *do_it_to_me_in_class* (female). As in this example, we generally observed an active/passive dichotomy between girls and boys.

Some girls' usernames were:

put_it_in_me_smoothly
fuck_me_wildly
make_me_feel_yours
visual_pleasure
do_it_to_me_in_public

Boys chose names like:

meet_me_in_my_bed
it_doesn't_fit_in_your_hand
single_and_no_commitment
I'll_set_you_on_fire
I've_got_what_you_want
night_without_a_condom

Pictures

Many pictures uploaded by girls and boys are compositions or doctored pictures. Very often faces are the most important or only element in the image (38% for girls and 33% for boys), which makes it difficult to know where the picture was taken. The rest of the pictures are usually taken somewhere in the home. Teenagers make their pictures in the bathroom (9% for girls, 17% for boys), in their bedroom (11%) or in other rooms in the home (13%). This means that many of these images reveal an intimate space in which mirrors are key (especially in the bathroom). Significantly, 23% of the girls and 14% of the boys upload pictures taken in public spaces.

There are very few pictures in which the body does not have an intended role: the vast majority of teenagers in Fotolog "pose" for the camera to a greater or lesser degree (90% for girls and 92% for boys). Both girls and boys appear with sexual outfits (40% and 41% respectively), but they show different parts of the body: 15% of the girls show off cleavage, whereas 28% of the boys show off muscles.

Although most subjects look into the camera when they take pictures of themselves (58% for girls and 45% for boys), we also found that many look into the void (15% and 16% respectively), and that girls appear bent forwards or lying on a bed or sofa more often than boys (25% versus 14%).

Boys and girls therefore choose different ways of portraying themselves. Unsurprisingly, these differences have to do mainly with aspects of sexual and gender identity: girls reproduce certain elements previously described by Mulvey (1999), Berger (2004) or Goffman (1979): our sample shows, for example, that girls portray their bodies in ways that invite the look of others (Mulvey's *male gaze*). They promote themselves with an explicit sexualisation of their bodies: as objects to be observed and, presumably, to be manipulated. They use ready-made descriptions of themselves in other (mainstream) media, assuming traditional looks and roles. They willingly present their bodies to the eye of the camera, presumably a male gaze. These self-representations are not far from those found in teen magazines, usually taken in intimate spaces and subtly – or sometimes not so subtly – showing key parts of the body like legs, cleavage or lower belly. Another example is in pictures where girls look into the void, as they usually reproduce what Goffman (1979) defines as *the distant woman*. This concept helps us to understand the body language in a large number of images in which girls, as opposed to boys who prefer to look at a point beyond the camera, portray themselves as being "lost in thought", distant and unreachable.

We also found a large number of other representations that could be considered as a part of the specific codes and patterns of Fotolog. This is the case for what we have called the *lesbian pose*: girls kissing, caressing or hugging each other. Lesbian pose may also imply a certain degree of action and autonomy regarding traditional representations, but the question remains whether this game is a genuinely critical appropriation of the feminine body and homo-sexuality, a transgression of imposed hetero-normativity, or, on the contrary, an action informed by the parameters of the *male gaze*. Are these images of two women in intimate poses aimed at pleasing men?

The self-presentations of boys express a subculture in which they perform and reproduce the "powerful man" concept in a peculiar style. On the one hand, in boys' images, their "look" is the *leitmotif*. Boys, more than girls, tend to portray themselves in clothes and haircuts that identify them as members of a specific group. On the other hand, they represent themselves as the Show-off (*chulo*) or the Naughty Boy. The naughty boy, for example, shows himself in provocative poses, as a conceited, tough boy. His picture emphasises attributes of power, especially muscles. That is what makes the equation of the *chulo*: if you are a Show-off you are attractive, and to be attractive, you must be a Show-off.

Texts

In 38% of the female posts, the text that accompanies the entry's picture is related to love and/or relationships. Although in 15 posts these texts refer to a positive relationship and 7 are declarations of love, the predominant message is the fear of commitment (31 posts). The most frequent concept of love is the one that associates love and pain (12 posts), and the one that considers love as inevitable (9 posts).

For boys, 43% of their texts are related to love. Positive relationships are described in 5 posts and declarations of love in 8 posts. The underlying concept of love in boys' texts is not generally clear.

A qualitative analysis of the texts shows that the traditional model of love is reproduced in the Fotolog environment. From the different examples that we found, four aspects emerged: a) the link between passion and pain (including suffering caused by violence), b) love as unpredictable and inevitable, c) dependence and d) negative feelings such as jealousy and resentment.

Passion/violence:

I love you above anything, above anything that crosses my mind. I love you when you insult me and you push me because you are angry with me. I love you when after that you start crying, you hug me and you tell me I'm your whole life (F25, female).

Being by his side is like riding an amazing roller coaster. When it comes down, it hurts and it makes me dizzy. When it goes up, is absolutely unstoppable. Over time I don't know if I prefer the way up or the way down (commentator, female).

Love and destiny:

We try to find happiness every day and we don't realise that it is happiness that will find us. And this will be when you least expect it: at school or at the supermarket, or in the middle of the sea (commentator, female).

Dependence and self-sacrifice:

Ask me whatever you want me to be: funny if you want, or pensive, intelligent or superstitious, courageous or even a dancer. I will be what you want me to be. Tell me and I will be that for you.

You're silly

I could be that […]

(F15, female).

I'm a junky of your voice […], my stomach hurts every time I feel that I am losing you… My greatest despair is that I don't want to get over being addicted to you. I want to die with you. I want your body, your smell, your voice, your kisses to kill me. (F313, female).

Negative feelings:

Let me make you suffer, so that you are the one to cry and suffer for having done what you did, for having abandoned me in the middle of that empty space that was so difficult to get out of, that hole I felt inside me because you were missing… now I am laughing about everything, after everything that happened (commentator, female).

It hurt me when I saw him talking to another girl, I was playing it cool, but inside I felt how jealousy was stinging like a thorn. Then he would come back, smile at me, kiss me, and all the thorns would disappear (commentator, female).

Thus, love is associated with "sparks" and chemistry, and hence understood as sudden and inevitable, and never as a conscious and responsible choice. Youngsters describe relationships as based on dependence and jealousy, acknowledging their suffering in dealing with these relationships. But they do not seem to look for a transformation or a change. Rather, they act on intuition, in terms of destiny, impulses and whims, more than in terms of reason. Passion, fantasy, dreams and magic are associated with the unknown and the inaccessible (Gómez, 2004).

Nevertheless, it must be said that in some cases passion and tenderness are linked (11 female posts and 9 male posts), and that youngsters also appreciate a boy or girlfriend who treats them right (5 female, 6 male). Finally, there are expressions of mutual support and cooperation (4 female, 5 male), concepts that have been related to the alternative model of love.

Comments

Real interaction between Fotolog users takes place in the Comments section. Based on the collected data, we cannot confirm that youngsters' main aim is to get as many comments as possible, but qualitative text analysis of the comments section shows that users often ask their readers to comment on their posts. Many of them also open the comments section with a self-comment in order to stimulate friends to participate. Other comments are strategic messages posted on a friend's blog in order to convince them to visit one's own blog: *come on and visit me* or *comment on me*.

The comments are usually very short phrases with a predominantly positive message (negative comments were found in only 2% of girls' logs and 3% of the boys'). In this sense, we think that social networks like Fotolog may strengthen friendships by means of positive comments, although it is clear that these are part of a *netiquette* in which positive messages are the rule. In the comments section there is no place for serious or profound topics, and contents uploaded by users is rarely questioned. The interaction on Fotolog can therefore hardly be defined in terms of dialogue but rather in terms of "I've seen your message and I'm here".

As to how youngsters address each other in updates or comments, we observed that girls call their female friends *my lesbian*, *my wife* or *my lover*. Girls play with these words, but it seems more like a pose or a provocation than a real transgression. For example: lesbians for life, hey people today is for my lesbian and me, lesbian I love you, my fiancée etc. Intimacy is redefined in two ways: on the one hand, user passwords were occasionally exchanged and used by friends to update each other's Fotolog, and on the other hand, one and the same user account covered several friends sharing a password or a nickname (*LEEEES 4 MEEM-BRIIIS!*, *HERMANAS*). Often, girls refer to each other as *sisters* or *twins*.

59% of all comments received by girls are positive support messages: phrases like *I love you* and *ILY loads* are mainly exchanged amongst girls, although girls also post them on boys' entries. Amongst girls, these affective tokens are occasionally mixed with swear words (*ILY bitch* or *ILY stupid*), or with sexual references (*ILY lover*, *ILY my little hooker*). Of all comments received by boys, 49% are this kind of positive message. Both male and female Fotolog users express declarations of eternal friendship to each other:

> You're really very important to me and you know that it's forever, don't you? I love you loads, baby, you're worth so much! (comment between girlfriends)

> Thanks for listening to me baby you are really top, I mean it (comment between girlfriends)

> You will always have me by your side, even though I am sometimes rude, you silly… I love you too much, more than you think and it is true that we are twin souls (comment between girlfriends)

21% of all positive comments received by girls refer to physical features. The words *good-looking* and *sexy* are the most frequently used. In boys' comments, this kind of messages represents 30% of all comments received.

Traces of a Traditional Model

After crossing the variables in image and text, we found a relationship between sexy outfits, low-cut necklines or exposed legs, and texts that refer to wanting to change partners or to not having worked hard enough for a relationship to work. The girls who show off their cleavage usually pose alone or with objects of beauty or power. Scenes of night-life usually include bare legs or otherwise erotically loaded outfits.

Teenagers who portray themselves together with their partner usually appear with them alone and have a positive attitude, as they express a link between love and stability, passion and tenderness in their texts. Moreover, they often express gratitude for the support and respect received from their partner.

Frequent portrayal with friends from both sexes in a nightlife setting is a typical feature of pre-adolescence. As opposed to the images of partners, in such pictures, where teenagers appear with friends or alone, there are no references to love being inevitable or to passion being associated to tenderness. The traditional model was thus confirmed, as we found a strong positive correlation between the stereotypes about love as something inevitable, love as painful and wanting somebody who is unreachable. We have also found texts referring to the desire to start or maintain a relationship with someone despite or (sometimes due to) the fact that it causes pain. This link between love as inevitable and the pain caused by it is confirmed by the text analysis above.

In the posts where an alternative model emerges, the correlation between love and stability, as well as between passion and tenderness is very clear. These posts also express gratitude for good treatment more explicitly. The conception of eternal love is associated with both the traditional and alternative models.

Finally, it is worth noting that teenagers, when they are rejected or when their relationship is in danger, strongly believe they will be able to change their partner ("I love him/her no matter what happens in the future, because I will change him/her").

Conclusions

A quantitative and qualitative analysis of 400 Spanish Fotolog usernames and entries (picture, text and comments) between September 2009 and February 2010 shows that teenagers, when representing themselves by means of self-generated media, mainly reproduce a traditional model of relationships. At worst, these representations suggest that certain violent traits can be attractive, especially to women. The reproduction of the traditional model of love and relationships in Fotolog emerges from the usernames. As we have seen, one out of five youngsters chose a name that explicitly refers to sex. Usernames of boys are generally aimed at stressing their sexual activity, whereas those of girls insinuate sexual availability.

We also found this active/passive dichotomy in the pictures, where boys pose as the *powerful and independent man*, while girls pose as the *passive and distant woman*. Remarkable was the *lesbian pose* in which girls appear hugging or kissing other girls. According to Mulvey (1999), both feminine and masculine subjects internalise a *male gaze*, which could explain why women experience pleasure when displaying themselves in situations that are (or are supposed to be) pleasing to men. In any case, the image of the body is never neutral or unintended: both girls and boys consciously pose for the camera, whether with friends and peers, or alone in the bathroom or bedroom.

The texts that accompany their pictures confirm the traditional model underlying their ideas about love and relationships. We distinguished four categories which all bear the traces of a traditional model: the link between passion and violence not only present in SNS but also in mainstream media representations, love as inevitable and uncontrollable, the link between love and negative feelings like jealousy, dependence and self-sacrifice. One of the most remarkable features in our analysis was that girls were the ones who expressed being dependent on their boyfriends. This seems to be the only tangible difference between boys and girls in Fotolog practices.

Although some boys also express dependence on their girlfriends, the vast majority adopts an independent and strong pose. As Giddens (Giddens, 1992) affirms, in the romantic ethos, boys who are desirable are precisely the cold and unreachable ones. That seems to be why boys represent themselves this way in their Fotolog entries, and why many of those representations effectively yield approving messages from female users. As opposed to this, women are enthusiastic and professional narrators of their personal lives, reproducing the pattern of the *quest-romance*, in which sexual encounters are seen as "detours on the way to an eventual love relationship" (Giddens, 1992, pp. 54-55). But in this quest, girls do not always represent themselves in passive ways: it is a task of active emotional engagement aimed at getting, maintaining, or getting back the person they love or want – a man located within the hetero-patriarchal rules of the traditional model. The final aim of their quest is a love relationship with a man after a process that usually entails pain and suffering, "worth bearing" for real love.

This *active* pose seems to be aimed at pleasing men (or at least a male gaze), so as to become a desirable sex object, as it is assumed that women should play an active role in sexual activities. Being this sex object is what seems to make feel girls powerful and accepted in their virtual peer groups. Our analysis shows that in the domain of love women are still socialised differently from men, which means that the former carry most of the emotional burden in order to sustain the relationship. As Giddens asserts, women's identity is strongly based on needs of others (Giddens, 1992, p. 85), which may explain why women are more prone than men to put up with abuse and violence, regardless of its intensity.

Despite these slightly disappointing results, however, our analysis also shows progressive features concerning attraction and choice of models of teenagers. For example, we found some girls expressing that they felt attracted by their partners because of their communication skills, and showing their gratitude for the support or respect received from their partners. In the same way, a few boys also demonstrated a respectful and egalitarian way of relating to girls. Moreover, in many comments from friends we found examples of peer solidarity, notably among women, which contradicts the same-sex competitiveness of the traditional model. Finally, we found clear references to the compatibility of passion and stability, as teenagers expressed their desire for an ever-lasting stable relationship with their current partners.

These progressive traits correspond to the depictions of Giddens of a new society where a *confluent love* is possible: active and contingent, and hence not ethereal (Giddens, 1992, p. 61). This love then leads to a "pure relationship" in which both subjects establish a non-utilitarian bond based on dialogue and mutual support.

After having analysed the risks but also the opportunities of Fotolog portrayals about love, sex and gender roles, we want to underline the importance of SNS in the identity-building and socialisation processes of youngsters. As a new medium, social networks allow youngsters to be in constant contact with each other at the same time. Fotolog shows us how teenagers appropriate narratives about relationships and express these from their own point of view, sharing them with a peer community and potentially with the whole world (being a public SNS). The conclusions of this study should draw the attention of educators, researchers and media makers to the continuous relationship between media messages and the affective socialisation of teenagers, both on an individual level and on the level of their discourse.

This analysis is an attempt to gain more insight into the causes of gender violence in Spain and beyond, and must serve to improve preventive measures with a view to building better and fairer relationships between young people by promoting cultural practices that do not associate passion with violence and pain, and by putting into place measures of preventive socialisation and education that do not sustain oppressive notions of love.

References

Altheide, L. & Snow, B. (1979). *Media Logic*. Beverly Hills, Calif.: Sage Publications.

Beck, U. & Beck-Gernsheim, E. (1998). *El normal caos del amor*. Barcelona: El Roure.

Berger, J. (2004). *Modos de ver*. Barcelona: Gustavo Gili.

Boyd, D. (2007). Why Youth (Heart) Social Network Sites: The Role of Networked Publics in Teenage Social Life. In D. Buckingham (ed.), *MacArthur Foundation Series on Digital Learning – Youth, Identity and Digital Media Volume* (pp. 119-142). Cambridge, MA: MIT Press.

Boyd, D.M. & Ellison, N.B. (2007). Social Network Sites: Definition, History, and Scholarship. *Journal of Computer-Mediated Communication, 13*(1). Retrieved May 19, 2011, from http://jcmc.indiana.edu/vol13/issue1/boyd.ellison.html.

Bruns, A. (2005). "Axel Bruns at iDC", Institute for Distributed Creativity. Retrieved May 19, 2011, from distributedcreativity.typepad.com/idc_events/2005/09/axel_bruns_at_i.html.

de Botton, L. & Oliver, E. (2009). Teoría crítica del radical love. *Revista Electrónica Teoría de la Educación. Educación y Cultura en la Sociedad de la Información, 10* (3), 90-102.

Duque, E. (2006). *Aprendiendo para el amor o para la violencia. Las relaciones en las discotecas*. Barcelona: El Roure.

Galician, M.L. (2004). *Sex, Love, and Romance in the Mass Media: Analysis and Criticism of Unrealistic Portrayals and Their Influence*. Mahwah, NJ and London: Lawrence Erlbaum Associates.

Giddens, A. (1995). *Modernidad e identidad del yo*. Barcelona: Península.

Giddens, A. (1992). *La transformación de la intimidad. Sexualidad, amor y erotismo en las sociedades modernas*. Madrid: Cátedra.

Gill, R. (2009). Mediated Intimacy and Post-feminism: A Discourse Analytic Examination of Sex and Relationships Advice in a Women's Magazine. *Discourse and Communication, 3*(4), 345-369.

Goffman, E. (1979). *Gender Advertisements*. New York: Harper and Row.

Goffman, E. (1991). La ritualización de la feminidad. In E. Goffman, *Los momentos y sus hombres* (pp. 135-168). Barcelona: Paidós.

Gómez, J. (2004). *El amor en la sociedad del riesgo. Una tentativa educativa*, Barcelona: El Roure.

Huffaker, D. & Calvert, S. (2005). Gender, Identity, and Language Use in Teenage Blogs. *Journal of Computer-Mediated Communication, 10*(2). Retrieved May 19, 2011, from http://jcmc.indiana.edu/vol10/issue2/huffaker.html.

L'Engle, K.L., Brown, J.D., & Kenneavy, K. (2006). The Mass Media Are an Important Context for Adolescents' Sexual Behavior. *Journal of Adolescent Health, 38*, 186-192.

Livingstone, S. (2009). Las redes sociales online, una oportunidad con riesgo para los adolescentes. In M. Grané & C. Willem (eds.), *Web 2.0: Nuevas formas de aprender y participar* (pp. 87-106), Barcelona: Laertes.

McLuhan, M. & Nevitt, B. (1972). *Take Today: The Executive as Dropout*. New York: Harcourt.

Mulvey, L. (1999). Visual Pleasure and Narrative Cinema. In L. Braudy & M. Cohen, *Film Theory and Criticism: Introductory Readings* (pp. 833-844). New York: Oxford University Press.

Oliver, E. & Valls, R. (2004). *Violencia de género. Investigaciones sobre quiénes, por qué y cómo superarla*. Barcelona: El Roure.

Padrós, M. (2007). *Models d'atracció dels i de les adolescents. Contribucions des de la socialització preventiva de la violència de gènere*. Institut Català de la Dona: Generalitat de Catalunya.

Ringrose, J. (2010). Sluts, Whores, Fat-Slags and Play-boy Bunnies: Teen Girls' Negotiations of "Sexy" on Social Networking Sites and at School. In C. Jackson, C. Paechter, & E. Renold (eds.), *Girls and Education. Continuing Concerns, New Agendas* (pp. 170-182). Basingstoke: Open University Press.

Sevick Bortree, D. (2005). Presentation of Self on the Web: An Ethnographic Study of Teenage Girls' Weblogs. *Education, Communication and Information*, *5*(1), 25-39.

Sveningsson, E.M. (2007). Doing and Undoing Gender in a Swedish Internet Community. In E.M. Seveningsson & J. Sundén (eds.), *Cyberfeminism in Northern Lights. Gender and Digital Media in a Nordic Context* (pp. 104-129). Cambridge: Cambridge Scholars Publishing.

Tapscott, D. (1996). *The Digital Economy: Promise and Peril in the Age of Networked Intelligence*. New York: McGraw-Hill.

Thelwall, M. (2008). Social Networks, Gender and Friending: An Analysis of MySpace Member Profiles. *Journal of the American Society for Information Science and Technology*, *59*(8), 1321-1330.

Valkenburg, P. & Peter, J. (2009). The Effects of Instant Messaging on the Quality of Adolescents' Existing Friendships: A Longitudinal Study. *Journal of Communication*, *59*(1), 79-97.

Valls, P., Puigvert, L., & Duque, E. (2008). Gender Violence Among Teenagers. *Violence Against Women*, *14*(7), 759-785.

Part IV

Empowerment and Protection

Parents, Teachers and Policy Makers' Role

Institutional Filters on Children's Internet Use

An Additional Explanation of Cross-national Differences in Parental Mediation

Veronika KALMUS & Triin ROOSALU[1]

Introduction

Along with the rapid growth of children's Internet use and rising public concern about risks and negative experiences kids may face online, there has been increasing research interest in parental strategies for regulating and monitoring children's online behaviour (see e.g. Kirwil (2009) for overview). Special Eurobarometer surveys in 2005 and 2008 have provided data on parental mediation of children's Internet use in EU 25 and EU 27 countries, respectively, allowing researchers to conduct comparative analyses, and leading to the overall conclusion that, in addition to individual-level variation in parental strategies, systematic cross-national differences exist (see Kalmus, Keller, & Pruulmann-Vengerfeldt, 2009; Kirwil, 2009; Kirwil, Garmendia, Garitaonandia, & Martínez-Fernández, 2009; Livingstone & Haddon, 2009; Lobe, Segers, & Tsaliki, 2009). These differences have mainly been explained by taking into account the countries' orientations in terms of individualistic and collectivistic values. This kind of approach is in line with a long tradition of research on techniques and practices of child-rearing guided by parental values and attitudes, which, in turn, are influenced by broader cultural ideologies (see Tulviste, Mizera, De Geer, & Tryggvason, 2007 for an overview).

[1] The preparation of this article was supported by the research grants No. 6892, 6968 and 8527 financed by the Estonian Science Foundation; the target financed projects No. 0180017s07, 0132682s05 and 0130027s11 from Estonian Ministry of Education and Science; the European Social Fund cofunding in EDUKO sub-project No. 30.2-10.2/1276, in Archimedes mobility grants No. 30.1-9.2/34 and 30.1-9.1/620 and in the Doctoral Schools program; a grant from the European Science Foundation (contract ref: FK/CE/SCSS-LD6-28), including support for research exchange visits; and Cornell University Institute for European Studies where Triin Roosalu was a visiting researcher when preparing the paper.

While this explanation is rather convincing, it is necessary to take into account that globalisation, along with the diffusion of new media technologies, by making parental experiences and perceptions more similar, has increased uncertainty and flexibility among parents in all countries (cf. Hofmeister, Blossfeld, & Mills, 2006). In the overall set of child-rearing activities, parental mediation of Internet use is a new phenomenon, and the current generation of parents is the first to face this challenge. They cannot rely entirely on their own childhood experiences or on the nation-specific cultural history, with its related meanings, implicit in the way child-rearing is interpreted in many other everyday contexts and practices. Hence, in acknowledging that the current generation of parents across countries is in a similar situation in having to work out strategies to mediate their children's Internet use, we find it intriguing that parents in different countries systematically prefer specific strategies.

In proceeding from the approach of Hofmeister and colleagues (2006), we assume that the universalising impact of globalisation on parental experiences and practices is mediated by institutional filters. In line with Spilerman (2009), we also suggest that country-level differences in value preferences affect the course of institutional adaptations to the unfolding of globalisation, both in rate and in the form that these adjustments will take. Referring to the differences between countries as a matter of institutional arrangements, as well as a matter of cultural mores and religious belief systems, provides the opportunity to disentangle the mechanisms behind these differences. We follow the concept of layered institutions, as presented by Scott (1994), who suggested these be viewed as consisting of three distinct elements: meaning systems and related behaviour patterns (e.g. time spent on different activities that form parenting); symbolic elements, including normative components (e.g. the prevailing vision in a society of a good parent and a well-behaved child); and regulatory processes for enforcement (informal and formal sanctions).

In this paper, we remain within the socialisation approach, which contextualises parental practices in relation to socialisation cultures. We assume, however, that the factors behind cross-national differences in parental strategies of mediation are multilateral. We suggest, firstly, that, besides individualistic and collectivistic values, other cultural factors deserve to be taken into consideration. Also, we take into account the level of Internet use among EU parents, as we assume it to be a necessary precondition for applying technical restrictions. More importantly, we seek to provide an additional explanation for cross-national differences in parental mediation by taking into account some institutional arrangements – namely, gender regimes (Hofäcker, 2006; Hofmeister &

Blossfeld, 2006) embedded – or reflected – in welfare state typologies (Esping-Andersen, 1990). We assume that the extent to which parents mediate their children's Internet use is influenced, among other factors, by the distribution of child-rearing tasks between the private and the public sphere, as well by the predominant gender role models in a given country.

In the following section, we give an overview of the relevant institutional arrangements, present a classification of European countries based on welfare and gender regimes, and outline our argumentation on how these institutional arrangements might affect parenting practices. In the third section, we discuss the available research on parental mediation of children's Internet use. Then we introduce our data and methodology. In the empirical analysis, by using cluster analysis of recent pan-European survey data, we provide a typology of parents based on their strategies of mediating children's Internet use, offer a classification of European countries according to the predominant parental types, and compare the results with the welfare state arrangements in a country to find out whether any systematic correlation patterns exist between the institutional arrangements and the daily practices that individual parents engage in. The paper concludes with a discussion focusing, among other issues, on policy implications, as well as on the way broader cultural ideologies influence the family as an agent of socialisation.

Institutional Arrangements Influencing the Distribution of Child-rearing Tasks

Based on a general normative assumption throughout the European cultural and geographical area and beyond, regardless of the specific gender regime, women tend to bear most of the domestic care-giving responsibilities, while men are more engaged in paid labour. The gender-based specialisation in home-making, although it has changed over time, ensures that there is always a parent at home to tend the children. This traditional pattern, however, is not dominant in all countries. The well-known welfare state typologies (Esping-Andersen, 1990), with ideas originating from classical approaches (see Holmwood, 2000; van Oorschot, Opielka, & Pfau-Effinger, 2008), have been systematically reformulated into different gender regimes (Esping-Andersen, Gallie, Hemerijck, & Myles, 2002; Blossfeld & Hofmeister, 2006). Pfau-Effinger (2004) has developed the concept of gender cultures as intertwined, but distinct, from gender regimes.

Analysing women's labour market participation in Western European countries and post-socialist states, Hofäcker (2006) has found the changes and developments in this regard to be compatible with the welfare regime of a specific state. The gender regime models applied in this type of analysis have as their key feature the extent to which the family structure invokes women as housewives or breadwinners (e.g. Bielenski, Bosch, & Wagner, 2002), and define differences based not only on norms but also on the form of the welfare state (see Walby, 2004). We use Hofäcker's (2006) classification as the starting point to differentiate between European countries, distinguishing between five types of institutional arrangements, here ordered according to the level to which parents are supported by private or public institutions in providing childcare (see Esping-Andersen, 1990; Esping-Andersen *et al.*, 2002; Folbre, 2008):

- *Southern European familialistic states* (such as Greece, Italy, Portugal and Spain) where neither the state nor the private sector have created any preconditions for the increased supply of female labour, as there are no alternative childcare opportunities besides the (extended) family. The predominant gender arrangement is the male breadwinner (see Bielenski *et al.*, 2002), with only limited part-time job opportunities for women.
- *Conservative states* are in the middle position in regard to women's labour force participation rates, as well as the role of the state and the private sector in increasing the supply of the female work force. In this cluster, Austria and Germany follow the predominant male breadwinner ideology, with rather high part-time opportunities for women, while France and the Netherlands fall into the modified male breadwinner model (Bielenski *et al.*, 2002).
- *Liberal states* (such as Ireland and the United Kingdom) have active labour force participation by women – not because of state support but due to the development of market-oriented childcare facilities and the service sector as an employer. The gender regime is characterised by the predominant male breadwinner pattern (Bielenski *et al.*, 2002), with wide part-time job opportunities.
- In *social democratic states* (such as Denmark, Finland, Norway and Sweden), women's labour market participation is very high, encouraged by state policies (e.g. creating public sector jobs and offering public childcare opportunities). The universal breadwinner model characterises these countries (Bielenski *et al.*, 2002). Part-time work is widely available, and women's part-time hours are longer than elsewhere.
- *Post-socialist states* are classified as a group, due to the fact that, following the societal transition to capitalism, the institutional systems have not yet stabilised enough to allow a more distinctive categorisation. Moreover, Hofäcker (2006) analysed only a small number of post-socialist countries and he did not reach any specific conclusions regarding the labour market participation rate of women.

> Some authors claim that post-socialist countries have already developed in different directions: a neo-liberal type in the Baltic states, an embedded neo-liberal type in the Visegrad states and a neo-corporatist type in Slovenia (Bohle & Greskovits, 2007). According to gender ideology, these countries can be labelled as following the universal breadwinner model, with very few part-time options and, thus, rather egalitarian full-time labour market participation, with mothers taking extensive breaks from their careers to care for younger babies (see also Roosalu & Täht, 2010).

On what basis do we assume that institutional arrangements such as welfare and gender regimes might help to explain the pattern of parental strategies of mediating children's Internet use? We assume that the extent and character of child-rearing practices, among them mediating children's online activities, depend on the amount of parental attention available to kids at home. This attention is probably highest in countries with predominant traditional role division between parents and lowest in regimes of two-career families.

At least two factors support this claim. First and foremost, the time effect is at work: the more parents are engaged in paid work, the less time they can probably devote to children. Thus, one would expect that when working parents' child-rearing practices are time-limited, they either pay less attention to this activity, or develop more efficient strategies. However, recent studies have demonstrated that this relationship is not entirely observable: while mothers are more often employed full-time than before, the overall time families devote to their children has increased across countries (Gauthier, Smeeding, & Furstenberg, 2004), as working parents decrease the time spent on leisure rather than on parenting. This may reflect the widespread belief in quality-time, or the understanding that it is not the amount of time a parent spends with children that matters, but the quality of activities one is engaged in with kids. Nevertheless, in the context of our comparative study, it is important to note that Gauthier and colleagues (2004) have witnessed significant differences between countries regarding the time spent directly on parenting, which may be attributable to the differences in parenting time that already existed before the global societal and cultural trends emerged.

Secondly, in countries with good provision of public childcare opportunities, parents may more easily outsource socialising tasks, including media education, to kindergartens and schools, thus taking a more passive role in mediating children's Internet use. Moreover, preschoolers educated at home may spend a considerable amount of time online, while children who attend kindergartens are less active Internet users, at least during the childcare hours. Thus, parents in countries with high childcare attendance may worry less about their children's media consumption.

An additional factor to be considered is the level of specialisation of child-rearing in a family, which may work in two different ways. In two-career families, both parents are engaged in paid work, as well as in parenting, leaving fewer personal resources, such as empathy and specific child-rearing experiences and skills, to be devoted to socialising kids. In male breadwinner or modified male breadwinner families, there is always one parent responsible for children's development and well-being, and this specialisation may help to develop the relevant skills. On the other hand, if both parents are engaged in employment as well as child-rearing, the family may have a wider variety of knowledge available to perform different parenting tasks. Mediating the use of technology, as a traditionally male-dominated sphere, may be one of those areas which are considered to be a responsibility of fathers. Also, working mothers may employ their professional resources (e.g. pedagogical or computer skills) in mediating their children's Internet use.

Thus, institutional arrangements embedded in different welfare and gender regimes, by influencing parents' time resources, child-rearing knowledge and skills, and allocation of socialising tasks between themselves and public institutions, probably also have some bearing on the predominant types of parental mediation of children's online behaviour.

Parental Strategies of Mediating Children's Internet Use

Previous literature offers several distinctions between general strategies parents use in mediating their children's Internet use, differentiating, for instance, between *system-based* and *user-based* approaches (i.e. between technical solutions and parental guidance), or *restrictive mediation* and *instructive mediation* (i.e. between rule-making and active efforts to interpret media content for children; see Kirwil *et al.*, 2009 for an overview). Based on these distinctions, various typologies of parental mediation of children's Internet use have been proposed. Lwin, Stanaland and Miyazaki (2008) propose four parental strategies: *restrictive*, *promotive* (only instructive mediation), *selective* (both restrictive and instructive), and *laissez faire* (no mediation). Livingstone and Helsper (2008) have described four factors of parental mediation: an active "co-use" and three types of "restrictive mediation" (use of technical filtering/monitoring tools, rule-making and monitoring of visited websites and e-mails).

Previous studies, almost unanimously, suggest that parents tend to favour the *user-based* or social approach over *system-based* or technical solutions (see e.g. Livingstone & Helsper, 2008; Lwin *et al.*, 2008; Kirwil, 2009; Kirwil *et al.*, 2009). Several individual-level differences between parents based on their gender, education and Internet use, as well as on the child's gender and age, have also been described in these studies. For instance, Kirwil and colleagues (2009) have shown that the more parents use the Internet, the more they practice social mediation and apply restrictions (with the exception of parents who use the Internet daily). The authors, however, do not explore the relationship between parents' Internet use and "system-based" mediation.

The existing cross-European comparative analysis suggests, in broad terms, that parents from individualistically-oriented child-rearing cultures (e.g. historically Protestant Nordic Europe) engage more in social co-use and apply more non-restrictive rules, and technical and website restrictions, while parents from cultures with a collectivistic orientation (e.g. Portugal and post-communist Europe, excluding Slovenia) use more time restriction (Kirwil 2009). This analysis was conducted by clustering eighteen European countries into four groups according to parents' preference for teaching individualistic or collectivistic values to children at home (based on the European Values Study 2000), and then comparing the distribution of five parental strategies of mediation in the four groups of countries. This type of analysis, however, does not account for the different combinations of strategies followed by parents; also, it clusters the countries into broad groups based on only one cultural dimension.

Data and Method

Our analysis is based on the survey data from *Flash Eurobarometer No. 248 – Safe Internet for children*, conducted in October 2008 among parents of 6-17-year-old children in 27 EU member states (N=12,803). We selected parents whose children accessed the Internet at home (N=8,631). Based on earlier distinctions between general strategies parents follow in mediating their children's Internet use (Kirwil, 2009; Kirwil *et al.*, 2009; Livingstone & Helsper, 2008; Lwin *et al.*, 2008), we composed four sum indexes of parental mediation:

- *Social mediation* (staying nearby when the child is online; sitting with the child when s/he goes online; asking/talking to the child about what s/he is doing or has done online; "always" added two points to the index, while "very frequently" added one point);
- *Monitoring mediation* (checking the computer later to see which sites the child has visited; checking the messages in the child's e-mail account/instant messaging service; checking whether the child has a profile on a social networking site/online community; "always" added two points to the index, and "very frequently" added one point);

- *Restrictive mediation* (not allowing the child to spend a lot of time online, to talk to people they don't know in real life, to use e-mail/ instant messaging tools, to use chat rooms, to create a profile in an online community, to access certain websites, to download/play music, films or games, to buy online; to give out personal information; each restriction added one point to the index);
- *Technical solutions* (filtering software; monitoring software; each solution added one point to the index).

We used the four indexes as input variables for two-step cluster analysis. A five-cluster solution, shown in Figure 4.1, provided the most comprehensive and most easily interpretable typology of parents according to their strategies for mediating children's Internet use.

Figure 1: Typology of parents according to their strategies for mediating children's Internet use (differences in the indexes compared to the average of the sample)

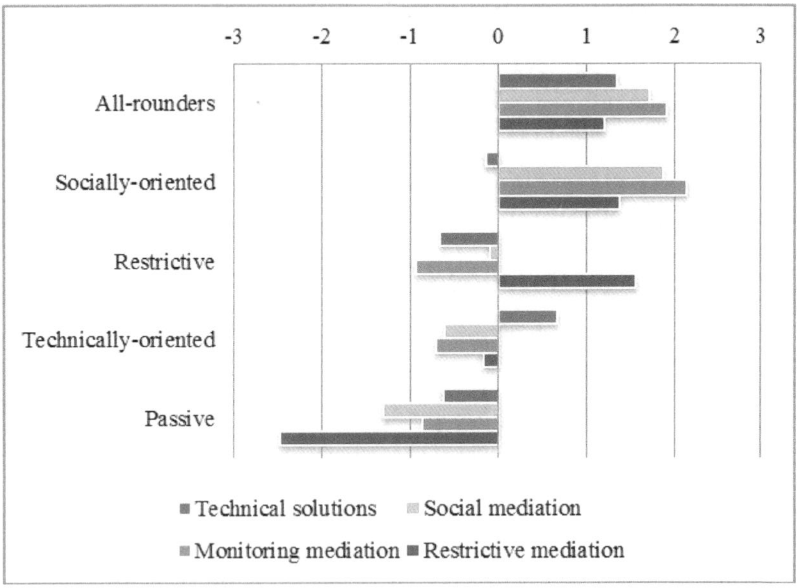

- *All-rounders* use all four types of mediation more actively compared to the average of the sample. This type is not widespread, involving only 10 per cent of European parents.
- *Socially-oriented* parents employ all "user-based" types of mediation most actively and make up 19 per cent of the sample.
- *Restrictive* parents, also comprising 19 per cent of the respondents, rely most heavily on making rules and setting restrictions.

- *Technically-oriented* parents use filtering and monitoring software relatively actively while practising "user-based" types of mediation less frequently than the average. The largest proportion of parents, 27 per cent, fell into the "technically-oriented" type.
- The remaining 25 per cent of parents practice all types of mediation, especially setting restrictions, less frequently compared to the sample average, and can be characterised as *passive* in mediating their children's Internet use.

To create a typology of EU member states, we cross-tabulated the types of parents with countries and classified the countries by predominant parental types, as shown in Table 1. We juxtaposed this classification with the level of parents' Internet use (derived from the same survey data) and the typology of welfare and gender regimes, based on Hofäcker's approach (2006).

Classification of Countries

The first group of countries (see Table 1) is characterized by an almost equal predominance of *all-rounders* and *technically-oriented* parents. The three countries in this group – the UK, Ireland and Germany – include two rather different large parental types: those who actively employ all possible means to mediate their children's Internet use, and those who rely mostly on technical solutions. The underlying welfare regimes (liberalism in the UK and Ireland, and conservatism in Germany), by providing wide part-time job opportunities but predominantly private childcare facilities, lead to differentiation of families between the male breadwinner type and two-career families, which is reflected in the two main strategies parents use to regulate children's online behaviour. As both of these strategies are based on parents' knowledge of the Internet, the high level of parents' Internet use in the UK and the medium level in Ireland and Germany can be seen as a precondition for the predominance of *all-rounders* and *technically-oriented* parents. In the UK and Ireland, parents ascribe both individualistic and collectivistic values in child-rearing moderate importance; in Germany, individualistic values also have moderate importance, while collectivistic values have low importance (Kirwil, 2009).

The second group of countries is distinguished by the predominance of *socially-oriented* parents, and includes only southern European familistic countries. With women's low labour force participation and modest availability of public childcare, children spend their time more regularly with an adult nearby, which also obviously facilitates social interaction with a parent when the child goes online. In Greece, Italy and Spain, individualistic values in child-rearing have moderate importance and collectivistic values have low importance, while Portugal belongs to the group of countries with moderate importance of collectivistic values and very low importance of individualistic values (Kirwil, 2009).

Table 1: Classification of EU 27 countries according to parental types (%) and welfare regimes

Predominant parental types	*Country*	*Welfare Regime*	*Parents' Internet use**	*All-rounders*	*Socially oriented*	*Restrictive*	*Technically oriented*	*Passive*
All-rounders & Technically-oriented	UK	Liberal	H	*34*	15	7	39	6
	Ireland	Liberal	M	*27*	24	16	30	4
	Germany	Conservative	M	*27*	24	12	27	10
Socially-oriented	Portugal	Familistic	M	8	45	*23*	11	13
	Greece	Familistic	L	10	39	17	*24*	10
	Italy	Familistic	M	7	38	*22*	20	12
	Cyprus	Familistic**	L	5	37	*21*	*21*	15
	Spain	Familistic	L	18	31	*23*	17	11
	Malta	Familistic**	L	7	28	18	*25*	23
Restrictive & Socially-oriented	Romania	Post-socialist	L	3	*28*	32	10	27
Technically-oriented & Restrictive	Poland	Post-socialist	M	8	21	*26*	29	16
	Slovenia	Post-socialist	M	10	14	*25*	28	23
Technically-oriented	France	Conservative	H	15	*16*	11	43	*16*
	Luxembourg	Conservative**	H	15	*16*	12	41	*16*
	Belgium	Conservative	H	7	*21*	17	37	18
	Austria	Conservative	M	13	*19*	19	33	16
	Netherlands	Conservative/ Social democratic***	H	10	18	19	33	*20*
Technically-oriented & Passive	Sweden	Social democratic	H	5	5	13	43	*34*
	Finland	Social democratic	H	9	12	22	33	*25*
	Denmark	Social democratic	H	4	6	24	*29*	38
	Hungary	Post-socialist	M	4	13	22	*28*	33
Passive	Estonia	Post-socialist	H	4	11	15	*22*	48
	Czech Republic	Post-socialist	H	2	14	19	*19*	46
	Latvia	Post-socialist	M	5	15	12	*22*	46
	Slovakia	Post-socialist	M	2	17	*19*	18	44
	Bulgaria	Post-socialist	H	3	*25*	21	14	38
	Lithuania	Post-socialist	M	2	9	*31*	22	36
			Total	10	19	19	27	25

* H – high (at least 93 per cent of parents use the Internet at least once a month); M – medium (86-92 per cent of parents use the Internet at least once a month); L – low (less than 86 per cent of parents use the Internet at least once a month). ** Classification suggestion by the authors. *** The Netherlands has, due to its peculiarities, sometimes been classified into social democratic cluster (Vis *et al.*, 2008).

Romania makes up a distinctive case, characterised by a very high proportion of parents practising the restrictive strategy, on the one hand, and a great share of *socially-oriented* parents, on the other hand. Different cultural and institutional factors may play a role here. Among post-socialist countries, Romania is the one with the lowest attendance of children in kindergarten (Roosalu & Täht, 2010) and with the lowest female labour force participation. Thus, parental supervision in childcare in general, and in mediating Internet use in particular may be more feasible and normalised as a part of parental responsibilities. The fact that Romania has one of the lowest levels of parents' Internet use among the EU countries may account for it having the smallest overall proportion of *all-rounders* and *technically-oriented* parents.

Two other post-socialist countries, Poland and Slovenia, form the fourth group, which is characterised by the prevalence of *technically-oriented* and *restrictive* parents. Again, several cultural and institutional factors probably are at play here. The two countries are different in regard to individualistic versus collectivistic orientation, with moderate importance of collectivistic values and very low importance of individualistic values in Poland, and moderate importance of individualistic values and low importance of collectivistic values in Slovenia (Kirwil, 2009). However, Poland and Slovenia share a Catholic religious background and Socialist past with Romania (and Lithuania), which may provide some explanation for the high proportion of *restrictive* parents in all four countries, as the *threatened values* (Padilla-Walker & Thompson, 2005), such as innocence and proper behaviour of children, are more important in these cultures. Also, Poland is referred to as the "strongest case for a return to the male breadwinner model" among post-socialist countries (Zhelyazkova & Valentova, 2009), being characterised by low kindergarten attendance and low female labour force participation, which, in turn, may account for its similarity with some Western European conservative countries (e.g. Austria and Germany) in terms of the share of *socially-oriented* and *technically-oriented* parents.

In the fifth group of countries, the greatest proportion of parents rely mostly on technical solutions in regulating their children's Internet use, while *socially-oriented* and/or *passive* parents also make up a large percentage. The group includes mostly conservative continental European countries. We can assume that the differentiating effect of welfare and gender regimes is rather similar to the one suggested in the case of the first group of countries, fostering extensive social mediation in families with male breadwinners and leaving two-career families with the options of technical solutions or passive strategies. In terms of individualistic versus collectivistic orientation, three different groups of countries are represented in this cluster. In Belgium, parents ascribe both individualistic and collectivistic values moderate importance; in France and Austria, individualistic values have moderate importance while collectivistic values have low importance; and the Netherlands stands out for the very high importance of individualistic values and very low importance of collectivistic values (Kirwil, 2009).

The sixth group of countries is characterised by relatively high percentages of *technically-oriented* and *passive* parents, and includes all three Nordic social democratic states, as well as post-socialist Hungary. Social democratic and post-socialist countries are known for women's active, full-time participation in the labour market, encouraged by state policies. The prevalent universal breadwinner model in these countries provides some explanation for the fact that most families opt for strategies which involve less supervision and social interaction when children go online. Also, the relative passivity of parents may be partly due to the higher level of outsourcing of child-rearing tasks to the state via the extensive use of public childcare beginning at age three. In this cluster, Sweden and Denmark are two of the countries with very high importance of individualistic values and very low importance of collectivistic values (Kirwil, 2009).

The seventh group of countries stands out as having the highest proportion of *passive* parents and is comprised of only post-socialist states. Compared to social democratic regimes, in post-socialist countries high female labour force participation is combined with very low part-time options, which may leave the parents with the least time and energy to employ more active strategies in mediating their children's online activities. In this group, Estonia, the Czech Republic and Bulgaria are characterised by moderate importance of collectivistic values and very low importance of individualistic values (Kirwil, 2009).

Discussion

The results of the comparative analysis confirm our assumption about the interplay of multilateral factors behind cross-national differences in parental mediation. Firstly, our analysis confirmed that as a precondition of the predominance of *all-rounders* and/or *technically-oriented* parents, at least a medium level of parents' Internet use is needed; however, a high level of parents' Internet use in itself is not sufficient for these two parental types to prevail in a country.

Secondly, our typological analysis, in general, lends support to the claim by Kirwil and colleagues that parental mediation of children's Internet use varies between the countries with an individualistic orientation in child-rearing and the countries where collectivistic values are more important (Kirwil, 2009; Kirwil *et al.*, 2009). Our results suggest, however, that the division of European countries into broad groups based on a single cultural dimension is a bit too wide a generalisation. For instance, Portugal as a collectivistic country firmly belongs, according to our analysis, in the same group as other southern European familistic countries, most of them characterised as individualistic. Also, post-socialist countries, most of them sharing a collectivistic orientation, vary greatly in terms of parental strategies. Thus, other cultural factors, among them religious background, need to be taken into account.

Thirdly, our analysis suggests that systematic correlation patterns indeed exist between the types of welfare arrangement (and underlying gender regimes) and predominant parental styles of mediating children's Internet use. It is therefore possible to suggest that welfare state institutions, especially through regulating female labour force participation and the availability of public childcare, have some bearing on the strategies parents are actually able to employ in their child-rearing tasks.

As a limitation of our study, we admit that the factors considered in our analysis are not sufficient to explain the full variety of parental strategies in European countries. Additional cultural and institutional indicators, such as the length and intensity of awareness-raising activities are needed to clear up further particular characteristics, for instance a great proportion of *all-rounders* in some countries.

Our findings, nevertheless, allow us to comment on the debate between cultural determinists and institutionalists, such as Lück (2006) and Spilerman (2009), on the role culture plays in the way societies are organised. Lück (2006) has argued that the cultural background of societies has shaped their institutional arrangements, but cultural mores and religious belief systems continue to assert a direct effect on women's values. He also claims that, compared with cultural background, the impact of the welfare regime is modest. According to Lück, institutional arrangements should not matter as much as culture in explaining the individual level practices. Spilerman (2009) is critical of the fact that studies of values tend to be oblivious to the literature of institutional effects. While Spilerman agrees with Lück that the cultural background of a society preceded and has shaped its institutional features, continuously constraining and moulding the institutional change, he suggests that the impact of value structures on individual-level practices is filtered through institutions.

In relying on the conclusions from the study by Kirwil and colleagues (2009) and on our own research (both based on the same Flash Eurobarometer survey data), we would like to emphasise that the influence of broader cultural ideologies and trends, both global and national, on families as agents of socialisation is, with a high degree of probability, both direct as well as filtered through institutional arrangements. Acknowledging the basic differences in the institutional arrangements across countries, and at the same time the standardising and unifying global forces, we also propose a further, more general question: How likely it is that the values and meanings – cultural framings – will be diverse in the future in different countries? One of the major mechanisms for inter-generational transmission of these kinds of framings is the family itself as an agent of socialisation, while regime differences largely predict predominant parenting styles in a country. The extent of the persistence of country-level differences in raising future generations and mediating their participation in the global (youth) culture and information flows accessible through online media thus leads us to predict the sustainability of national differences regardless of the cultural and technological trends towards homogenisation.

As a policy implication, stemming from an analysis of the relation between parental strategies of mediation and their effectiveness in terms of reducing children's experience of online risks, researchers boldly suggest that it is "the task of policy makers to promote parental mediation of Internet use, particularly in countries with a collectivistic culture in which parental mediation tends to be lower" (Kirwil *et al.*, 2009, p. 213). However, parents should be guided in their strategies according to the given socialisation culture, as the same strategy may function as either a protective factor against risk or one that increases it, depending on the cultural context (Kirwil *et al.*, 2009; Lobe *et al.*, 2009). In addition, our findings clearly demonstrate the need for more attention to the effects of welfare arrangements in different countries, not least when planning for awareness-raising campaigns among parents. Also, when welfare regimes foster the outsourcing of socialising tasks to public childcare, media education deserves more emphasis in the curricula of kindergartens and primary schools.

References

Bielenski, H., Bosch, G., & Wagner, A. (2002). *Working Time Preferences in Sixteen European Countries*. Dublin: European Foundation for the Improvement of Living and Working Conditions.

Blossfeld, H.-P. & Hofmeister, H. (2006). *Globalization, Uncertainty and Women's Careers: An International Comparison*. Cheltenham, UK & Northampton, MA: Edward Elgar.

Bohle, D. & Greskovits, B. (2007). Neoliberalism, Embedded Neoliberalism and Neocorporatism: Towards Transnational Capitalism in Central-Eastern Europe. *West European Politics*, *30*(3), 443-466.

Esping-Andersen, G. (1990). *The Three Worlds of Welfare Capitalism*. Cambridge: Polity Press & Princeton: Princeton University Press.

Esping-Andersen, G., Gallie, D., Hemerijck, A., & Myles, J. (2002). *Why We Need a Welfare State*. Oxford: Oxford University Press.

Folbre, N. (2008). *Valuing Children: Rethinking the Economics of the Family*. Cambridge: Harvard University Press.

Gauthier, A.H., Smeeding, T.M., & Furstenberg, F.F. Jr. (2004). Are Parents Investing Less Time in Children? Trends in Selected Industrialized Countries. *Population and Development Review, 30*(4), 647-672.

Hofäcker, D. (2006). Women's Employment in Times of Globalization: A Comparative Overview. In H.-P. Blossfeld & H. Hofmeister (eds.), *Globalization, Uncertainty and Women's Careers: An International Comparison* (pp. 32-60). Cheltenham, UK & Northampton, MA: Edward Elgar.

Hofmeister, H. & Blossfeld, H.-P. (2006). Women's Careers in an Era of Uncertainty: Conclusions from a 13-Country International Comparison. In H.-P. Blossfeld & H. Hofmeister (eds.), *Globalization, Uncertainty and Women's Careers: An International Comparison* (pp. 433-450). Cheltenham, UK & Northampton, MA: Edward Elgar.

Hofmeister, H., Blossfeld, H.-P., & Mills, M. (2006). Globalization, Uncertainty and Women's Mid-Career Life Courses: A Theoretical Framework. In H.-P. Blossfeld & H. Hofmeister (eds.), *Globalization, Uncertainty and Women's Careers: An International Comparison* (pp. 3-31). Cheltenham, UK & Northampton, MA: Edward Elgar.

Holmwood, J. (2000). Three Pillars of Welfare State Theory: T.H. Marshall, Karl Polanyi and Alva Myrdal in Defence of the National Welfare State. *European Journal of Social Theory, 3*(23), 23-50.

Kalmus, V., Keller, M., & Pruulmann-Vengerfeldt, P. (2009). Children and Young People in the Consumer and Information Society. In M. Lauristin (ed.), *Estonian Human Development Report 2008* (pp. 115-122). Tallinn: Estonian Cooperation Assembly.

Kirwil, L. (2009). Parental Mediation of Children's Internet Use in Different European Countries. *Journal of Children and Media, 3*(4), 394-409.

Kirwil, L., Garmendia, M., Garitaonandia, C., & Martínez-Fernández, G. (2009). Parental Mediation. In S. Livingstone & L. Haddon (eds.), *Kids Online: Opportunities and Risks for Children* (pp. 199-215). Bristol: Policy Press.

Livingstone, S. & Haddon, L. (2009). *EU Kids Online: Final Report*. London: LSE, EU Kids Online (EC Safer Internet Plus Programme Deliverable D6.5). Retrieved April 29, 2010, from www.lse.ac.uk/collections/EUKidsOnline.

Livingstone, S. & Helsper, E.J. (2008). Parental Mediation of Children's Internet Use. *Journal of Broadcasting and Electronic Media, 52*(4), 581-599.

Lobe, B., Segers, K., & Tsaliki, L. (2009). The Role of Parental Mediation in Explaining Cross-National Experiences of Risk. In S. Livingstone & L. Haddon (eds.), *Kids Online: Opportunities and Risks for Children* (pp. 173-183). Bristol: Policy Press.

Lück, D. (2006). The Impact of Gender Role Attitudes on Women's Life Courses. In H.-P. Blossfeld & H. Hofmeister (eds.), *Globalization, Uncertainty and Women's Careers: An International Comparison* (pp. 405-432). Cheltenham, UK & Northampton, MA: Edward Elgar.

Lwin, M.O., Stanaland, A.J.S., & Miyazaki, A.D. (2008). Protecting Children's Privacy Online: How Parental Mediation Strategies Affect Website Safeguard Effectiveness. *Journal of Retailing, 84*(2), 205-217.

van Oorschot, W., Opielka, M., & Pfau-Effinger, B. (2008). *Culture and Welfare State: Values and Social Policy in Comparative Perspective.* Cheltenham: Edward Elgar Publishing.

Padilla-Walker, L.M. & Thompson, R.A. (2005). Combating Conflicting Messages of Values: A Closer Look at Parental Strategies. *Social Development, 14*(2), 305-323.

Pfau-Effinger, B. (2004). *Development of Culture, Welfare States and Women's Employment in Europe*. Aldershot: Ashgate Publishing.

Roosalu, T. & Täht, K. (2010). Women in the Labour Market: The Effect of Parental Leave on Career Opportunities. In M. Lauristin (ed.), *Estonian Human Development Report 2009* (pp. 96-99). Tallinn: Estonian Cooperation Assembly.

Scott, R.W. (1994). Institutions and Organisations: Towards a Theoretical Synthesis. In R.W. Scott & J.W. Meyer (eds.), *Institutional Environments and Organizations: Structural Complexity and Individualism* (pp. 55-80). Thousand Oaks: Sage.

Spilerman, S. (2009). How Globalisation Has Impacted Labour: A Review Essay. *European Sociological Review, 25*(1), 73-86.

Tulviste, T., Mizera, L., De Geer, B., & Tryggvason, M.T. (2007). Child-Rearing Goals of Estonian, Finnish, and Swedish Mothers. *Scandinavian Journal of Psychology, 48*, 487-497.

Vis, B., van Kersbergen, K., & Becker, U. (2008). The Politics of Welfare State Reform in the Netherlands: Explaining a Never-Ending Puzzle. *Acta Politica, 43*, 333–356.

Walby, S. (2004). The European Union and Gender Equality: Emergent Varieties of Gender Regime. *Social Politics, 11*(1), 4-29.

Zhelyazkova, N. & Valentova, M. (2009). Women's Perceptions of Consequences of Career Interruption Due to Childcare in Central and Eastern Europe. In *IRISS Working Paper 2009-01*. Luxembourg: CEPS/INSTEAD. Retrieved February 7, 2010, from iriss.ceps.lu/documents/irisswp99.pdf.

Alternative Regulatory Instruments Versus Fundamental Rights

Match Made in Heaven or Recipe for Disaster?

Eva LIEVENS

Introduction

The fundamental legitimacy of the traditional policy goal of protecting minors against harmful content has not been called into question by the developments in the media sector. Traditionally, with respect to "old" media, this aim of protecting minors against harmful content has been aspired to by means of legislation, for instance by imposing a watershed in a Media Act or by establishing film classification legislation. However, over the past decades, the use of conventional legislation – also dubbed "command-and-control regulation" – which entails that the state performs all regulatory tasks (creation, implementation and monitoring, and enforcement), increasingly displayed a number of shortcomings, especially in complex sectors such as the media sector. Examples of such shortcomings are the territoriality of traditional legislation, the slow-moving wheels of bureaucracy, and a lack of expertise and involvement of knowledgeable actors, for instance with regard to very specific issues caused by technological evolutions.

Hence, decentred forms of regulation, which are much more open to the involvement of different actors in the regulatory process, gained significant attention. This development also occurred with regard to the protection of minors against harmful content where enthusiasm for the use of alternative regulatory instruments (ARIs), such as self- and co-regulation, grew over the past decade.

Self-regulation entails the creation, implementation and enforcement of rules by a group of actors with no – or at least minimal – involvement of actors that do not belong to this group (such as the government). Codes of conduct established by a group of industry actors are an example of self-regulation. Whereas in the 1990s self-regulation was hailed as the panacea to regulate the digital media environment due to its flexibility, adaptability, higher degree of incorporated expertise and incentives for commitment and compliance, critics note that this regulatory instrument has a significant number of drawbacks such as a lack of effective enforcement, a low level of transparency and accountability, the fact that private interests are put before public interests and the unaccountability of private actors to the public.

Co-regulation is an instrument which consists of elements of state regulation and elements of self-regulation. The structure of co-regulatory mechanisms might vary. For instance, such mechanisms might consist of an initiative by the state (e.g. by enacting a legal basis), implementation by private actors and supervision again by the state.[1] As a "middle road" between the state and self-regulation, it has been argued that co-regulation combines the advantages of both these instruments (on the one hand, the flexibility, fast adaptation, expertise and engagement of the industry – assets of self-regulation, and, on the other hand, legal certainty, democratic guarantees and more efficient enforcement – benefits of legislation) while remedying their drawbacks. Especially with respect to important public policy goals, such as the protection of minors against harmful content, it is important to be able to fall back on a governmental backbone if private interests threaten to undermine them. Gunningham and Rees stated that "there is reason to believe that self-regulatory mechanisms under-pinned by some form of state intervention are more resilient and effective than self-regulation in isolation" (Gunningham & Rees, 1997, p. 366). The HBI/EMR Study on Co-regulation in the Media Sector (Hans-Bredow-Institut & EMR, 2006) also identified a number of advantages related to the cooperation of state and non-state regulation. According to the authors, co-regulatory systems have a chance of higher industry accountability, can benefit from the faster pace of decision-making and demonstrate a greater sustainability. However, it would not be correct to present co-regulation as a panacea without any drawbacks whatsoever. Prosser has pointed out that, in certain circumstances, co-regulatory systems do not offer the best of both worlds but the worst, "in which neither [private or public interests are] respected and any values are subjected to unprincipled bargaining between the state and private interests" (Prosser, 2008, p. 103). It is important to stress that co-regulatory systems need to be carefully drawn up. Normative requirements or *process values* (Prosser, 2008), such as transparency, adequate participation, and independence, need to be respected and the division of

1 An example can be found in the Netherlands, with respect to the *Kijkwijzer* system, which is a cross-media content classification system. *Kijkwijzer* aims to provide parents with age recommendations and information about the content of television programmes, movies shown in cinemas, DVDs, and mobile content. This system has a basis in the Dutch Media Act but the implementation is in the hands of a co-regulatory body ("NICAM") to which content providers are affiliated. The government monitors the functioning of the system and the Dutch Media Authority can regulate providers who are not affiliated to the co-regulatory body. For more information: www.kijkwijzer.nl.

tasks between the different actors needs to be clearly established. Accountability and credibility are crucial requirements for the non-state components of regulation. Furthermore, the exact combination of state and non-state elements will need to be structured carefully, so as not to lose the advantages of both self-regulation and traditional state regulation (Marsden, 2004).

Fundamental Rights

Fundamental or constitutional rights are the cornerstone of democratic legal systems. It is hence of the utmost importance that ARIs – which do not function within a legal vacuum, although their name may suggest otherwise – respect these rights. This is even more crucial in delicate matters, such as the protection of minors against harmful content, where a meticulous balance needs to be found between different goals of public interest.

Since the use of ARIs was first contemplated, warnings have been issued about the compatibility of such systems with certain fundamental rights. The delegation of regulatory power to non-state bodies, a key characteristic of most ARIs, where fundamental rights are – potentially – at stake has thus always been regarded with scepticism. In this context, it has been advocated that the state should remain the primary guarantor of human rights (Council of Europe, 2005; McGonagle, 2003). It can also be recalled that in the first documents broaching co-regulation, such as the White Paper on European Governance, the European Commission stressed that co-regulation should not be considered in cases where fundamental rights are called into question (European Commission, 2001). The rationale behind this was, of course, the concern that private actors should not determine the scope of regulations which could restrict fundamental rights. However, since the adoption of the White Paper in 2001, it has been argued that the unsuitability of the use of co-regulation in situations where fundamental rights are at stake should not be interpreted quite so strictly (Hans-Bredow-Institut & EMR, 2006; Rand Europe, 2007). Moreover, it has also been suggested that collaboration between the state, industry and civil society could result in a more complete protection of certain fundamental rights (Council of Europe, 2005).

The complex balancing exercise between ensuring an adequate level of protection of a number of fundamental rights, on the one hand, and taking advantage of the positive aspects of the use of ARIs to protect minors in a digital environment, on the other hand, is the subject of this chapter. One of the main issues that will be addressed is the fact that, traditionally, provisions which aim to guarantee fundamental rights are directed at public authorities or governments. Alternative regulatory strategies, however, often delegate responsibilities to alternative regulatory bodies. The key question which arises in such a situation is whether the legal safeguards with regard to the fundamental rights described above will still be applicable.

Four fundamental rights are especially relevant to the subject at hand: two *substantive* rights, the right to freedom of expression and the right to privacy, and two *procedural* rights, the right to a fair trial and the right to an effective remedy. In the form of a checklist, this chapter presents an analysis of the applicability of the safeguards contained in the relevant articles of the European Convention on Human Rights (ECHR) and of the requirements that need to be met to ensure an optimal protection when ARIs are used to protect minors from digital harmful content.

Substantive Human Rights and the Use of ARIs

Freedom of Expression (Article 10 ECHR)

A first substantive right is the right to freedom of expression, which is of the utmost importance with regard to the protection of minors from harmful content. In this context, conflicts may arise between two competing interests: on the one hand, the freedom of expression of adults who are allowed to legally access harmful material,[2] and, on the other hand, the protection of minors against such material, which a legitimate goal of public interest, recognised for instance by the European Court of Justice,[3] the European Court of Human Rights[4] and in article 17 of the United Nations Convention on the Rights of the Child (UNCRC). An added competing interest might be the freedom of expression of children, laid down in article 13 of the UNCRC.

Article 10 ECHR, the cornerstone of the European legal framework regarding freedom of expression, stipulates that everyone has the "freedom to hold opinions and to receive and impart information and ideas without interference by public authority and regardless of frontiers". This fundamental right, however, is not an absolute right. Restrictions, i.e., "*formalities, conditions, restrictions or penalties*",[5] such as the seizure of published material, criminal sanctions or labelling used to filter

---

2 Please note that this paper does not deal with illegal content, but focuses on harmful content, i.e. content which is legal for adults to access but which may be considered harmful for certain vulnerable groups such as children and young people.

3 ECJ, Dynamic Medien v. Avides Media AG, C-244/06, 14.02.2008, paras. 42 and 47.

4 European Court of Human Rights, Handyside v. the United Kingdom, 07.12.1976, para. 52.

5 Article 10 para. 2. ECHR.

content (Hans-Bredow-Institut & EMR, 2006), can be allowed, but only if certain – strictly interpreted (Council of Europe, 2007; van Dijk, van Hoof, van Rijn, & Zwaak, 2006; Voorhoof, 2004), – conditions are fulfilled.

Interferences with freedom of expression by public authorities thus need to be prescribed by law, must pursue a legitimate aim (such as the protection of morals, which is the aim most suitable in the field of protecting minors against harmful content), and must be necessary in a democratic society (article 10 para. 2 ECHR). Three main issues can be raised: first, whether the fact that article 10 ECHR targets interferences by "public authorities" results in the inapplicability of article 10 ECHR with respect to the use of ARIs, second, whether there are particular difficulties with respect to the three conditions of article 10 para. 2 ECHR, and third, whether there are specific issues related to the use of self-regulation and the involvement of private actors in the human rights context.

Is article 10 ECHR applicable when an infringement is committed by a self-regulatory or co-regulatory body (as opposed to a "public authority")?

When ARIs are used, the potential interference of the freedom of expression will often – depending on the structure of the ARIs in question – originate from a non-state body. Hence, the question is whether article 10 ECHR is at all applicable when restrictions are part of a self-regulatory or co-regulatory scheme. The answer to this question will depend on the level of government involvement in the scheme. If the restrictive measures do stem from a state actor (for instance, in a co-regulatory scheme), or from an actor which can be considered – according to the circumstances – as a "public authority", then the applicability of article 10 ECHR will not be questioned (Hans-Bredow-Institut & EMR 2006). In the case *Casado Coca v. Spain*, for instance, the European Court of Human Rights decided that the Barcelona Bar Council was, according to Spanish law, a public law corporation, and, moreover, that the purpose of the Bar Council served the public interest, and that the penalty imposed by the Bar Council had been upheld in national courts.[6] Hence, the Court decided that there was in fact interference by a public authority and, consequently, article 10 ECHR was considered applicable.[7] In the case of a purely self-regulatory scheme or *soft* co-regulatory scheme with a low degree of government involvement, however, this would probably not be the case.

6 European Court of Human Rights, Casado Coca v. Spain, 24.02.1994, para. 39.

7 *Ibid.*

At this point, a second question can be raised, i.e. to what extent the Convention may be applied to relations between private or non-state actors. According to the horizontal effect theory, in certain circumstances, individuals or private actors can – if the national legal system accepts the direct effect of the Convention articles – invoke article 10 ECHR before the national courts to challenge other individuals. Furthermore, even when this is not the case, the European Court of Human Rights has judged that, again in certain circumstances, a *positive obligation* may rest on public authorities to guarantee that the freedom of expression of their citizens is not infringed upon by private actors. In the *Fuentes Bobo v. Spain* case, for instance, the Court acknowledged that "a positive obligation can rest with the authorities to protect the freedom of expression against infringements, even by private persons".[8] Furthermore, in *VGT Verein v. Tierfabriken*, the Court stated that "in addition to the primarily negative undertaking of a State to abstain from interference in Convention guarantees, 'there may be positive obligations inherent' in such guarantees".[9] Hence, it could be argued that in certain circumstances public authorities should take measures to stop the interferences with the right to freedom of expression by self- or co-regulatory bodies.

It is very difficult, however, to make general assumptions about the application of horizontal effect with respect to ARIs, since there is no consensus on its scope (Van Dijk *et al.*, 2006). The European Court of Human Rights has even explicitly stated that it does not wish "to elaborate a general theory concerning the extent to which the Convention guarantees should be extended to relations between private individuals inter se".[10]

Can ARIs, which impose restrictions on the right to freedom of expression in order to protect minors from harmful digital content, fulfil the conditions of article 10 para. 2 ECHR?

The fulfilment of the condition that an interference with the right to freedom of expression must pursue a *legitimate aim* will in the majority of cases not pose significant problems, since the protection of minors is a legitimate aim which can fall under the "protection of morals" or the "protection of the rights of others".

With regard to the second requirement, it can be noted that restrictions imposed by self- or co-regulatory schemes will not always be *prescribed by a law*. However, the European Court of Human Rights has

8 European Court of Human Rights, Fuentes Bobo v. Spain, 29.02.2000.

9 European Court of Human Rights, VGT Verein v. Tierfabriken, 28.06.2001, para. 45.

10 European Court of Human Rights, VGT Verein v. Tierfabriken, 28.06.2001, para. 46.

not always required the legal basis for a restriction to be included in a law *stricto sensu*: other instruments can be attributed a law-like status on the condition that they are accessible and foreseeable. The *Barthold v. Germany* case clarified, for instance, that rules of professional conduct could be attributed such a status.[11] However, in that particular case, the level of government involvement was rather high. Hence, it is likely that the Court would not be as tolerant with respect to ARIs in which there is no or a very minimal level of government involvement.

To meet the third requirement, i.e. *necessity in a democratic society*, ARIs need to be proportional, the reasons for the justification of the restriction need to be relevant and sufficient, and there has to be a pressing social need. In order to assess this, the competing interests (the protection of minors against harmful content, and the protection of the freedom of expression) will need to be balanced on a case-by-case basis.

What is the impact of the use of (pure) self-regulation on the right to freedom of expression?

It is a fact that private actors or intermediaries, such as Internet Service Providers (ISPs), mobile telephony providers or search engine operators (Nikoltchev, 2008), play a far greater role in the new information and communication environment compared to the traditional media landscape, which was dominated by a much more limited number of broadcasters and newspaper publishers. Those private "new media" actors are often indispensable points of access to the new communication technologies and, hence, as the Council of Europe pointed out: "their role is a prerequisite for enabling and empowering users to access the benefits of the information society, in particular to seek and impart information and ideas, to create and to access knowledge and education" (Council of Europe, 2008). However, this also means that these private actors have the power to restrict access to the services they provide, for instance, by removing, filtering or blocking content. These restrictions carried out by private actors may thus have a significant impact on the freedom of expression.

One of the fears that has frequently been voiced over the past decade is that restrictions on the freedom to expression by private actors, for instance by means of filtering or blocking schemes, in fact amount to private censorship. As was mentioned above, according to article 10 ECHR, states or *public authorities* are prevented from interfering with the freedom of expression of their citizens if no serious justifications can be put forward. However, we have also found that the applicability of article 10 ECHR is not as straightforward with respect to private actors

[11] European Court of Human Rights, Barthold v. Germany, 25.03.1985.

(cf. *supra*: horizontal effect). Hence, if questionable actions by private actors do not fall within the remit of article 10 ECHR, the protection of fundamental rights is at stake. Schemes in which private actors carry the full responsibility for filtering, labelling or blocking content, for instance, to protect minors from harmful content, have therefore been regarded with scepticism by many commentators. Just as state censorship cannot be tolerated, nor should private censorship. Distinguishing between what is illegal and what is harmful is a very difficult and delicate issue that should not be left to private companies (Noorlander, 2003). In this context, the Council of Europe stated in its *Human Rights Guidelines for Internet Service Providers* that ISPs which provide access services, hosting, applications or content, should not be "expected to advise on what content or behaviours are illegal and/or harmful" (Council of Europe, 2008). Furthermore, with respect to ISPs which offer hosting, applications or content, the Council of Europe (2008, para. 21) stressed that

> [i]n respect of filtering, blocking or removal of illegal content, you should do so only after a verification of the illegality of the content, for instance by contacting the competent law enforcement authorities. Acting without first checking and verifying may be considered as an interference with legal content and with the rights and freedoms of those creating, communicating and accessing such content, in particular the right to freedom of expression and information.

The theory that ISPs should take the utmost care in dealing with alleged illegal content – which can actually be illegal, but can also be *simply* harmful – might, however, in practice, prove to be challenging. It is, for instance, unclear what exactly is meant by *contacting the competent law enforcement authorities*. How far should ISPs go to achieve a satisfactory answer to the question whether certain content is illegal? Most often, only a court will be able to decide upon the legality or illegality of certain content. Furthermore, taking the liability regime imposed by the EU e-Commerce Directive into account, the issue only becomes more complicated, since this regime requires, for instance, hosting providers who have an *actual knowledge of illegal content* to block access to that content. This can potentially result in a chilling effect, given that this regime encourages ISPs to err on the side of caution by taking down content which is not illegal (McIntyre & Scott, 2008).

At the very least, private companies who are involved in self-regulatory systems related to filtering and blocking harmful content, should employ transparent procedures and be very open and straight about the criteria they take into account when deciding upon the potential harmfulness of content. Again, we can refer to the Council of Europe Guidelines which specify that ISPs providing hosting, applications and content should "[m]ake sure that any filtering or blocking of services carried out is legitimate, proportional and transparent to [their] customers" (Council of Europe, 2008, para. 20). Furthermore, they should

> inform [their] customers of any filtering or blocking software installed on [their] servers that may lead to a removal or inaccessibility of content as well as the nature of filtering that takes place (form of filtering, general criteria used to filter, reasons for applying filters) (Council of Europe, 2008, para. 20).

In our view, given the dangers of private censorship, self-regulatory schemes in which only private actors are in control of removing, filtering or blocking content are not ideal to protect minors from harmful content. Especially with respect to such a delicate normative goal, balanced co-regulatory schemes in which the government,[12] and hence, the public interest, is represented to a greater extent, might provide more guarantees.

Privacy (Article 8 ECHR)

Children's privacy is often at risk in the digital information and communications environment (Walrave, 2005), for instance, when they are asked to transmit their personal details in exchange for a prize, or when identification and age verification tools are used to check their age before gaining access to a particular website. The right to privacy has been included in a number of legal provisions. A specific right for children has been affirmed by article 16 of the United Nations Convention on the Rights of the Child (Article 29 Data Protection Working Party, 2008 & 2009). At the European level, one of the most important foundations for the right to privacy is article 8 ECHR.[13] This article is similar to article 10 in that its second paragraph also offers *public authorities* the possibility of limiting the basic right to privacy, provided that certain conditions are fulfilled (i.e. such measures need to be in accordance with the law and necessary in a democratic society, as well as pursuing a legitimate interest, such as the protection of health or morals). In fact, when it comes to the relationship between ARIs and privacy, the questions that are raised are rather similar to the ones which surfaced with respect to article 10 ECHR: first, can article 8 ECHR be applied to acts of private actors (who are involved in the functioning of ARIs aimed at protecting minors against harmful content), and, second,

---

12 Without overlooking, of course, the requirements of article 10 para. 2 ECHR with respect to restrictions imposed on the freedom of expression by public authorities.

13 In addition, specific EU directives, such as the Data Protection Directive and the Directive on Privacy and Electronic Communications, relate to the right to privacy.

how is the reference to *the law* (as in the first condition of article 8 para. 2 ECHR) interpreted?

Is article 8 ECHR applicable when the infringement is committed by a self-regulatory or co-regulatory body (as opposed to a "public authority")?

The analysis regarding this potential application with respect to article 10 ECHR is also valid with respect to article 8 ECHR (*supra*). One additional note that could be made is that the European Court of Human Rights has often referred to the positive obligation states have to protect an individual's right to privacy. In *X. and Y. v. the Netherlands*, for instance, the Court stated:

> The Court recalls that although the object of Article 8 (art. 8) is essentially that of protecting the individual against arbitrary interference by the public authorities, it does not merely compel the State to abstain from such interference: in addition to this primarily negative undertaking, there may be positive obligations inherent in an effective respect for private or family life (see the Airey judgment of 9 October 1979, Series A No. 32, p. 17, para. 32). These obligations may involve the adoption of measures designed to secure respect for private life even in the sphere of the relations of individuals between themselves.[14]

The actual adoption of such measures by the state depends on an assessment of the reasonable balance that needs to be reached between the competing interests of the individual and the community, and on the margin of appreciation of the state (Kilkelly, 2001; Van Dijk *et al.*, 2006). In a recent case, *K.U. v. Finland*, the European Court of Human Rights specified, for instance, that notwithstanding this margin of appreciation, "effective deterrence against grave acts, where fundamental values and essential aspects of private life are at stake, requires efficient criminal-law provisions".[15] The case dealt with an advertisement on a dating site, placed by unknown persons, in the name of a 12-year-old boy without his knowledge. This advertisement included the age of the boy, a description of his physical characteristics, a link to his website which contained a picture and a telephone number, and a statement that he was seeking an intimate relationship with a boy.[16] At the time of the facts it was not possible according to Finnish legislation to obtain the identity of the person who placed the advertisement from the Internet provider.[17] The Court considered the applicability of article 8 ECHR

[14] European Court of Human Rights, X. and Y. v. the Netherlands, 26.03.1985, para. 23.

[15] European Court of Human Rights, K.U. v. Finland, 02.12.2008, paras. 42-43.

[16] European Court of Human Rights, K.U. v. Finland, 02.12.2008, para. 7.

[17] European Court of Human Rights, K.U. v. Finland, 02.12.2008, para. 40.

indisputable,[18] and emphasised that "[c]hildren and other vulnerable individuals are entitled to State protection, in the form of effective deterrence, from such grave types of interference with essential aspects of their private lives".[19] The fact that no effective steps could be taken to identify and prosecute the person who placed the advertisement, and thus the failure by the Finnish government to fulfil its positive obligation to provide a framework of protection, led the Court to decide that article 8 ECHR had been violated.[20] Although the Court's decision has raised some controversy, this case is interesting because it shows that the Court takes into account the vulnerability of children and the existence of positive obligations when their privacy is interfered with.

We can conclude that, depending on the circumstances, a certain horizontal effect *supra* of article 8 ECHR is conceivable (Van Dijk *et al.*, 2006).

Can ARIs, which impose restrictions on the right to privacy in order to protect minors from harmful digital content fulfil the conditions of article 8 para. 2 ECHR?

The analysis regarding this possibility with respect to freedom of expression is also valid with respect to the right to privacy (*supra*). This entails that, on a case-by-case basis, an assessment will need to be made whether the ARIs in question achieve a legitimate aim (*in casu* the "protection of morals" or the "protection of the rights of others"), can be considered to be "prescribed by law" (i.e. have "a legal basis", are "sufficiently precise and contain a measure of protection against arbitrariness by public authorities", and are "foreseeable"; Kilkelly, 2001, p. 25), and are necessary in a democratic society (i.e. there is a pressing social need for the interference with the right to privacy and the measures taken do not go beyond what is necessary to achieve the legitimate aim).

Procedural Human Rights and the Use of ARIs

Aside from *substantive rights*, certain *procedural rights* also need to be respected. Although their relevance might seems less obvious to the topic at hand, procedural rights are inextricably linked to the substantive rights which come into play when instruments are set up to protect minors from harmful content. When important rights, such as freedom of expression and privacy, are dealt with by specific regulatory instruments, adherence to certain procedural safeguards by the decision-making bodies, *in casu* self- or co-regulatory bodies, is essential. The

18 European Court of Human Rights, K.U. v. Finland, 02.12.2008, para. 41.

19 European Court of Human Rights, K.U. v. Finland, 02.12.2008, para. 46.

20 European Court of Human Rights, K.U. v. Finland, 02.12.2008, paras. 49-50.

right to a fair trial in case of a dispute and the right to an effective remedy in case of a violation of a person's rights, both laid down in the European Convention on Human Rights, are especially relevant within the context of the use of alternative regulatory instruments, and the status of the alternative regulatory bodies in particular.

Right to a Fair Trial (Article 6 ECHR)

A first right that might be of importance to the use of ARIs is the right to a fair trial, which implies that an individual has the right to a fair and public hearing before an independent and impartial tribunal established by law, when a dispute concerning this individual's (civil) rights occurs.

Is article 6 ECHR applicable with respect to the use of ARIs to protect minors against harmful content?

The applicability of article 6 ECHR is triggered by a dispute concerning an individual's civil rights. It has been accepted that the right to freedom of expression and the right to privacy – the two rights which will most often be interfered with when ARIs are used to protect minors from harmful content – may be considered civil rights. Hence, when such an ARI gives rise to a dispute concerning an individual's right to freedom of expression or privacy (or of course any other civil right), article 6 ECHR will be applicable. It is important to note that article 6 ECHR only comes into play when there is a dispute which is real and serious and which relates to an actual right.[21]

Can an alternative regulatory body be considered a "tribunal established by law"?

When it has been established that article 6 ECHR is applicable, it is necessary to examine whether recourse to an "alternative regulatory body" which exercises functions which are determinative of civil rights (Mole & Harby, 2006), might be considered as satisfying the requirement of a trial before a "tribunal established by law". This could be the case if the European Court of Human Rights deems that the body is question is characterised by its judicial function, i.e., that it determines matters within its competence on the basis of rules of law and after proceedings conducted in a prescribed manner. Furthermore, the Court will assess the body's independent and impartial character and will check whether other procedural guarantees are provided. In our view, it is highly unlikely that a self-regulatory body will be considered to meet these standards. It may even be rather doubtful with respect to co-

21 European Court of Human Rights, Van Marle and others v. the Netherlands, 26.06.1986, para. 32.

regulatory bodies. However, this will need to be assessed on a case-by-case basis.

Which procedural guarantees, included in article 6 ECHR, do which alternative regulatory bodies need to comply with?

There are two scenarios. First, if the alternative regulatory body is considered a "tribunal" then all procedural guarantees provided by article 6 ECHR (e.g., fair and public hearing, independence, impartiality, public judgment, reasonable time frame, etc.) need to be adhered to. However, if – because of the "non-classic" nature of the decision-making body – not all requirements are complied with, the doctrine of full review could be considered applicable. This entails that the Court will accept this situation provided that a full judicial review is available. Second, if the alternative regulatory body in question cannot be considered a tribunal or if there is simply no possibility provided within the ARI to deal with a dispute concerning an individual's civil rights, the individual in question should have access to a tribunal that does comply with all procedural guarantees that are contained in article 6 ECHR.

Right to an Effective Remedy Before a National Authority (Article 13 ECHR)

A final fundamental right that is relevant with respect to the use of ARIs, is laid down in article 13 ECHR. This right entails that anyone who claims that his or her rights or freedoms under the ECHR have been violated must have access to an effective remedy before a national authority.[22] Article 13 ECHR complements the other rights and freedoms guaranteed by the Convention. If an individual's rights are interfered with, for instance by a decision of a self- of co-regulatory body, the protection of these rights would be futile if no effective remedy is available.

Can remedies provided by an alternative regulatory body be considered as remedies before a "national authority"?

When it comes to ARIs, the question arises whether remedies provided by alternative regulatory bodies qualify as "remedies before a national authority". This could be the case if the European Court of Human Rights is of the opinion that the "authorities" in question are independent and impartial, and provide certain procedural guarantees. It can be noted that the Court does not require remedies to be judicial, nor national authorities to be judicial authorities in the strict sense. The

[22] It is not necessarily required that children themselves have a right to an effective remedy, as long as a legal representative of the child, who is involved in a violation of Convention rights, has recourse to a remedy.

authorities in question, however, need to be competent to receive a complaint, to investigate the merits of the complaint and to take binding decisions regarding the provision of redress.

When can remedies provided by an alternative regulatory body be considered effective?

In theory, the remedy should either lead to the prevention of the suspected violation or, if appropriate, to the obtainment of adequate redress, including compensation, and should be effective in practice as well as in law. This will be considered by the European Court of Human Rights on a case-by-case basis. However, case-law has already demonstrated that the standards of the Court are rather high.

In *Peck v. the United Kingdom*, the Court considered remedies provided by certain (alternative regulatory) bodies not effective because of their lack of legal power to award damages. In this case, the European Court of Human Rights found that a violation of article 8 ECHR had taken place when CCTV images of the applicant were made public through the media. The applicant had also claimed that article 13 was violated as, according to him, no remedy regarding this infringement of his right to privacy was available. The respondent (the UK government) argued that (aside from other remedies)[23] the applicant had been able to assert and vindicate his claims before the Broadcasting Standards Commission (BSC), the Independent Television Commission (ITC) and the Press Complaints Commission (PCC),[24] bodies which could be considered to be self- or co-regulatory (Hans-Bredow-Institut & EMR, 2006). The Court, however, decided that these bodies could not provide an effective remedy for the applicant:

> The Court finds that the lack of legal power of the commissions to award damages to the applicant means that those bodies could not provide an effective remedy to him. It notes that the ITC's power to impose a fine on the relevant television company does not amount to an award of damages to the applicant. While the applicant was aware of the Council's disclosures prior to the *Yellow Advertiser* article of February 1996 and the BBC broad-

23 European Court of Human Rights, Peck v. the United Kingdom, 28.01.2003, para. 95.

24 European Court of Human Rights, Peck v. the United Kingdom, 28.01.2003, para. 94. Regarding the status of the remedies provided by these bodies, the respondent stated that they "accepted that it was not intended that the media commissions should provide a 'legal remedy, in the sense of making pecuniary compensation available to an aggrieved individual who may have been injured by an infringement of the relevant codes'. However, they contended that article 13 did not require in every case 'a court' or that a pecuniary award be available" (para. 94).

casts, neither the BSC nor the PCC had the power to prevent such publications or broadcasts.[25]

Hence, a violation of article 13 was asserted by the Court.

From this judgment it could be deduced that – dependent, of course, on the concrete circumstances of each case – if alternative regulatory bodies are involved in the protection of Convention rights, such as the freedom of expression, in the framework of a system to protect minors from harmful content – the availability of remedies should be carefully considered (Craufurd Smith, 2006). Although the European Court of Human Rights has up until now not dealt with such cases, it would not be unimaginable that an adult, for instance, would claim that his or her freedom of expression has been restricted by a self- or co-regulatory scheme to protect minors against harmful content, and that he or she would seek to remedy this infringement by appealing to an alternative regulatory body. The fact that the European Court of Human Rights would not automatically assume that such bodies are not classifiable as *national authorities* is an important finding, and one that should be taken into account when self- and/or co-regulatory models are established that could conflict with the freedom of expression. However, it will, again, depend on the circumstances of the case whether the remedy provided will comply with the – rather high – standards used by the Court to assess the effectiveness of a remedy. If the Court considers that a remedy provided by a self- or co-regulatory body is not adequate, other remedies (such as for instance before the national courts) need to be available.

Should there be an effective remedy before a national authority when an infringement is committed by a private actor (for instance, in a self-regulatory scheme)?

It has been suggested that, in theory, the final part of article 13 ECHR ("notwithstanding that the violation has been committed by persons acting in an official capacity") could imply that an effective legal remedy must also be supplied when a violation has been committed by a private individual, according to the theory of indirect horizontal effect of the Convention rights between citizens. However, at the same time it has been argued that this should be confined to cases where a positive obligation exists to safeguard individuals against serious violations committed by other individuals, as, for instance, related to article 2 (right to life) or article 3 (protection from torture and other inhuman or degrading treatment or punishment) (Van Dijk *et al.*, 2006). Nevertheless, the

[25] European Court of Human Rights, Peck v. the United Kingdom, 28.01.2003, para. 109.

European Court of Human Rights did acknowledge in *Plattform Ärtze für das Leben v. Austria* that the applicability of article 13 was linked to article 11, which – according to the Court – "sometimes requires positive measures to be taken, even in the sphere of relations between individuals, if need be".[26] It is thus not inconceivable that in the context of article 10 ECHR (freedom of expression) or article 8 (right to privacy), the applicability of article 13 to relations between individuals would be considered possible as well.

Conclusions

The aim of this paper was to examine whether there are legal obstacles to the use of ARIs to protect minors against harmful content from a human rights point of view, and more specifically, from the point of view of a number of rights embedded in the ECHR. We distinguished two categories of fundamental rights which could be affected: first, the substantive rights to freedom of expression and privacy, and, second, the procedural rights to a fair trial and to an effective remedy.

With regard to the substantive fundamental rights (which are in theory addressed at states or governments) we can conclude that their applicability will often depend on the level of government involvement in ARIs.[27] This means that a number of provisions relating to the protection of the right to freedom of expression and the right to privacy will be more likely to apply when there is a degree of government involvement, as is common with respect to co-regulatory systems. Conversely, self-regulatory systems may fall outside of the protection of the legal framework (except, for instance, when theories, such as the *horizontal effect* theory can be applied). In our opinion, this might be dangerous in a delicate area such as the one that is at the centre of this chapter.[28] Hence, to protect minors from exposure to harmful digital content, the use of co-regulatory systems, where there is an actual symbiosis between the involvement of the government and other actors, and greater guarantees

---

26 European Court of Human Rights, Plattform Ärtze für das Leben v. Austria, 21.06.1988, para. 32.

27 With respect to the level of government involvement that is required in order for certain legal provisions to be applicable, it might be noted that, up until now, both the European Court of Human Rights and the European Court of Justice have been very rarely faced with cases in which measures similar to ARIs (to protect minors from harmful content) were at issue. It can be assumed that as more ARIs are used, the chance that the Courts will have the opportunity to bring greater clarity to this matter will increase.

28 We can frame this finding also within the current general "malaise" with respect to self-regulation or regulation by the market or the sector (cf. for instance, the financial crisis). As a consequence, in different sectors, the calls for a renewed and more intense involvement of the government have recently grown louder.

are provided as to the actual realisation of the policy objective, is preferable. If instruments which may harm the right to freedom of expression or the right to privacy fall within the protection of the human rights framework, restrictions on these rights can be monitored, assessed and remedied in a more consistent and efficient manner, and, consequently, the protection of minors will be enhanced.

With regard to the procedural guarantees, it can be argued that the compliance with such guarantees is not only essential for an ARI to be reliable; it also ensures that the rights that are at stake are suitably protected. Unfortunately, procedural rights are often overlooked when ARIs are set up. Yet, when important rights are at stake, when decisions are made that might interfere with such rights, the least that can be expected from a decision-making body is adherence to certain procedural safeguards, such as independence, impartiality, and transparency (e.g. by means of reasoned decisions). In addition, it is essential that decisions of alternative regulatory bodies can be disputed in an adequate manner.

It is not possible to overemphasise the significance of respecting substantive as well as procedural fundamental rights when trying to protect minors against harmful content. Finding the right balance between the competing values is very delicate and will therefore need to be carefully considered when establishing ARIs in this field, whatever their shape or form.

References

Article 29 Data Protection Working Party (2008). *Working Document 1/2008 on the Protection of Children's Personal Data (General Guidelines and the Special Case of Schools)*. Retrieved March 5, 2008, from http://ec.europa.eu/justice_home/fsj/privacy/docs/wpdocs/2008/wp147_en.pdf.

Article 29 Data Protection Working Party (2009). *Opinion 2/2009 on the Protection of Children's Personal Data (General Guidelines and the Special Case of Schools).* Retrieved May 18, 2009, from http://ec.europa.eu/justice_home/fsj/privacy/docs/wpdocs/2009/wp160_en.pdf.

Council of Europe (2005). *Pan-european Forum on Human Rights in the Information Society: Responsible Behaviour by Key Actors.* Retrieved April 25, 2007, from http://www.coe.int/t/dghl/standardsetting/media/doc/ForumStbgSept2005Report_en.asp#TopOfPage.

Council of Europe (2007). *Freedom of Expression in Europe: Case-law Concerning Article 10 of the European Convention on Human Rights.* Strasbourg: Council of Europe Publishing.

Council of Europe (2008). *Human Rights Guidelines for Internet Service Providers – Developed by the Council of Europe in Co-operation with the European Internet Service Providers Association (EuroISPA)*. Retrieved October 10, 2008, from http://www.coe.int/t/dghl/standardsetting/media/Doc/H-Inf(2008)009_en.pdf.

Craufurd Smith, R. (2006). European Community Media Regulation in a Converging Environment. In N. Shuibhne (ed.), *Regulating the Internal Market* (pp. 105-143). Cheltenham: Edward Elgar.

European Commission (2001). *European Governance – A White Paper, COM* (2001) 428 final.

Gunningham, N. & Rees, J. (1997). Industry Self-regulation: An Institutional Perspective. *Law & Policy, 19*(4), 370-380.

Hans-Bredow-Institut & EMR (2006). *Study on Co-regulation Measures in the Media Sector: Final Report, Study Commissioned by the European Commission.* Retrieved July 28, 2006, from http://ec.europa.eu/avpolicy/docs/library/studies/coregul/final_rep_en.pdf.

Kilkelly, U. (2001). *The Right to Respect for Private and Family Life: A Guide to the Implementation of Article 8 of the European Convention on Human Rights (Human Rights Handbook, No. 1)*. Strasbourg: Council of Europe.

Marsden, C. (2004). Co- and Self-regulation in European Media and Internet Sectors: The Results of Oxford University's Study www.selfregulation.info. In C. Möller & A. Amouroux (eds.), *The Media Freedom Internet Cookbook* (pp. 76-100).Vienna: OSCE.

McGonagle, T. (2003). The Potential for Practice of an Intangible Idea. In S. Nikoltchev (ed.), *Co-regulation of the Media in Europe* (pp. 2-8). Strasbourg: European Audiovisual Observatory.

McIntyre, T.J. & Scott, C. (2008). Internet Filtering: Rhetoric, Legitimacy, Accountability and Responsibility. In R. Brownsword & K. Yeung (eds.), *Regulating Technologies: Legal Futures, Regulatory Frames and Technological Fixes* (pp. 109-124). Oxford: Hart Publishing.

Mole, N. & Harby, C. (2006). *The Right to a Fair Trial: A Guide to the Implementation of Article 6 of the European Convention on Human Rights (Human Rights Handbooks, No. 3)*. Strasbourg: Council of Europe.

Nikoltchev, S. (ed.) (2008). *Searching for Audiovisual Content.* Strasbourg: European Audiovisual Observatory.

Noorlander, P. (2003). Freedom of Expression and Internet Regulation. In C. Hardy & C. Möller (eds.), *Spreading the Word on the Internet: 16 Answers to 4 Questions (Reflections on Freedom of the Media and the Internet Amsterdam Conference, June 2003)* (pp. 105-115). Vienna: OSCE. Retrieved April 23, 2007, from http://www.osce.org/publications/rfm/2003/06/12245_103_en.pdf.

Prosser, T. (2008). Self-regulation, Co-regulation and the Audio-Visual Media Services Directive. *Journal of Consumer Policy*, 31(1), 99-113.

Rand Europe (2007). *Options for Effectiveness of Internet Self- and Co-regulation – Inception Report, Study Commissioned by the European Commission.* Retrieved January 8, 2008, from http://ec.europa.eu/dgs/information_society/evaluation/data/pdf/studies/s2006_05/inception_final.pdf.

van Dijk, P., van Hoof, F., van Rijn, A., & Zwaak, L. (eds.) (2006). *Theory and Practice of the European Convention on Human Rights*. Antwerpen: Intersentia.

Voorhoof, D. (2004). Vrijheid van meningsuiting [*Freedom of expression*]. In J. Vande Lanotte & Y. Haeck, *Handboek EVRM: Deel II: Artikelsgewijze commentaar (Volume I)* [*ECHR: Part II: Commentary on the articles*] (pp. 837-1061). Antwerpen: Intersentia [in Dutch].

Walrave, M. (ed.) (2005). *Cyberkids' e-Privacy – Minderjarigen, minder rechten?* (Privacy Paper Nr. 4) [Kids' and Teens' e-Privacy at Stake? An Analysis of Data Processing and Privacy Statements on Websites Aimed at Children and Adolescents]. Retrieved 14 June, 2007, from www.e-privacy.be/PrivacyPaper4-Cyberkids-e-Privacy.pdf [in Dutch].

Regulatory Initiatives for Managing Online Risks and Opportunities for Youths

The East Asian Experience

Sun Sun LIM

Introduction

This study reviews policy interventions, non-governmental initiatives and industry responses pertaining to Internet safety for children in three East Asian countries – China, Japan and South Korea (hereafter, Korea). It will assess regional trends and pay special attention to policies and initiatives unique to each country. The three countries constitute an interesting slate of comparisons because of the significant growth in media and technology adoption in recent years and their socio-cultural characteristics. With information technology being widely deployed in business and education in the three countries, children's knowledge of computers and the Internet is regarded as critical to academic success and social advancement (Lim, 2008). At the same time, China, Japan and Korea share a communitarian, Confucianist background with a pronounced protective and controlling approach towards raising children, and a strong emphasis on academic achievement. Consequently, children are the focus of considerable parental and societal investment. This study examines how this confluence of economic, cultural and social factors has influenced the nature and direction of policy interventions, non-governmental initiatives and industry responses *vis-à-vis* children's Internet safety in the three countries.

This study is based on 21 expert interviews with 47 representatives of regulatory bodies, public and private media companies, non-governmental organisations, civic groups and academia in China, Japan and South Korea (see Appendix). The interviews were conducted in 2007 as part of a fact-finding mission of a Singaporean non-governmental advisory body – the Advisory Council on the Social Impact of New Media on Society (AIMS). AIMS comprised media industry veterans, civil servants and academics who had been appointed by the government on the basis of their experience and expertise in the media.

AIMS had been tasked by Singapore's Minister for Information, Communication and the Arts to prepare a report[1] on the social and societal implications of Singapore's avid adoption of new media, and to suggest ways of managing these implications. The author was a member of the AIMS Working Group and was part of the fact-finding mission which visited China, Japan and Korea to seek a deeper understanding of how these countries were managing the social impact of new media. The interviews covered a range of topics including the impact of the Internet on state-citizen communication, the regulation of online political content and the protection of minors. This article will focus only on the interview discussions pertaining to the protection of minors.[2] It will first discuss cultural values and conceptions of children in the three East Asian countries, before reviewing salient Internet adoption trends, especially amongst the young. It will then explicate the three countries' key concerns about promoting children's Internet safety and their corresponding policy interventions, non-governmental initiatives and industry responses, while highlighting how these have been shaped by a confluence of social, economic and cultural factors. The key findings emerging from the interviews are then analysed with reference to a review of relevant secondary literature including academic research, official reports and media articles. Finally, it will conclude with a discussion of the limitations of the study and directions for future research.

Cultural Values and Conceptions of Children in East Asia

Politically, the three East Asian countries studied in this paper have different systems, with China being the sole communist country, while Japan and Korea are functioning democracies. Yet, they share a communitarian, Confucian background that has historically stressed respect for authority and hierarchy, community rights before individual rights, obedience and industry (Tu, 1996). Similarly, beliefs about childhood in the three countries can be traced to Confucian and Buddhist views of children and have historically influenced the nature of parenting in these countries (Chao & Tseng, 2002). Both Confucian and Buddhist worldviews share the notion that the child is intrinsically pure and innocent, and can only be corrupted by adult influences (Chao & Tseng, 2002). It is believed that until and unless they are tainted by negative influences

1 The complete report can be accessed here http://app.mica.gov.sg/Default.aspx?tabid=129.

2 All of the interviews were conducted between a group of working group members (including myself) and a group of officials from the host countries' agencies and corporations. Since these officials were speaking in their capacity as representatives of their respective organisations, no permission was sought to quote them directly or personally through verbatim quotes. Hence, their views are presented in indirect, reported speech.

in their environment, children are naturally good. Consequently, childhood is considered a precious and momentous period in one's life where parental involvement in guiding, nurturing and protecting the child is critical.

In China, besides shielding children from adverse environmental influences, it is believed that they need to be protected from their own impulses as they may not fully understand the consequences of their own actions (Ho, 1986). According to Chinese tradition, mothers are duty bound to control (*guan*) their children and to train them to rein in their impulses (Ho, 1986). Parental control is associated not with negative but positive connotations and is regarded as an expression of love and care (Wu, 1996b). Traditionally, Chinese children are socialised to be obedient (*tinghua*) and to defer to parental authority (Gao, Ting-Toomey & Gudykunst, 1996; Wu, 1996a). Similarly, the Japanese feel that the key to a child's success in life is a hierarchical parent-child relationship where children are obedient and diligent, and parents are instructive and nurturing (White, 1993). In their socio-emotional development, Japanese children are encouraged to be more sensible and to behave like someone who is a year older (Chen, 1996). Light corporal punishment and the modelling of good behaviour by parents are used to instil self-control in children (Hendry, 1986; Machida, 1996). Likewise, Korean parenting also emphasizes nurture, self-discipline and the use of corporal punishment (Kim, Kim, & Rue, 1997). The Korean family remains a hierarchical one, where deference to parental authority, although less strict and formalized than before, is still expressed by children and expected by mothers (MacDonald, 1996).

In addition to these traditional beliefs about the significance of children and childhood, social trends in the three countries have amplified the importance of children and influenced the nature of childrearing. All three countries have seen declining birth-rates, thus resulting in smaller families with fewer children. In China, while its overall birth-rate remains high, the strict enforcement of the one-child policy, particularly in the urban areas, has resulted in one-child families becoming the norm (Gu, 1990). In Korea, modernisation, urbanisation and industrialisation have contributed to a decline in the fertility rate, from 6.0 in 1960 to 1.19 in 2004 (Kim, Park, Kwon, & Koo, 2005). This trend is paralleled in Japan, where the fertility rate fell from around 5 in the 1920s to 1.29 in 2003 (Holloway, Suzuki, Yamamoto, & Mindnich, 2006). With fewer children to care for, parents in the three countries have become even more interested in investing in the future of their offspring (Chang, 2007; Chee, 2000; Hendry, 2003). Across the three countries, it is perhaps in education that the intensity of parental investment in their children is most palpable. The valorisation of academic achievement in these countries also has its roots in Confucianism which sees education as the key to honing one's inner virtues and to serving society productively (Chao & Tseng, 2002). Chinese culture has long emphasized the value of education, believed to be critical for one's social mobility and personal development (Hau & Salili, 1996; Stevenson & Lee, 1996). The one-child policy has in turn exacerbated the amount of pressure placed on only-children to perform academically (Chen, Lee, & Stevenson, 1996). Chinese families thus place great importance on education and are prepared to spend a large proportion of their family income to raise their children's educational level through extra tuition and enrichment classes (Anagnost, 1997; Wang & Cui, 2004). Likewise, Japanese society places great emphasis on education and the prestige of one's school (educational credentialism – *gakureishakai*) (Chen, 1996, p. 39). A good education is regarded as the key to success in life (Vaughn, 1996) and as with their Chinese counterparts, Japanese parents are avidly sending their children to enrichment classes to enhance their learning opportunities (Chen, 1996). Similarly, for Koreans, the child's academic success is a barometer of the family's status (Connor, 2002). Parents therefore allocate a large proportion of their total household expenditure to their children's education, to the detriment of their own retirement funds (Lee & Huh, 2004). Entrance into a top university is considered an integral step towards social and professional advancement and enrolling in a *hakwon* (cram school) is thought to increase one's chances (Cotton & Leest, 1996). In the face of such societal and familial pressures, it is unsurprising therefore that Korean adolescents are themselves achievement-oriented and tend to lead time-pressed lifestyles (Kim, 1994). In all three countries, effort, diligence and industry are seen as being integral to academic achievement (Chao & Tseng, 2002; Chen & Stevenson, 1995).

Internet Adoption Trends in East Asia

The governments of China, Japan and Korea have invested heavily in the development of Internet, digital and mobile communications infrastructure (Bleha, 2005; Lee, O'Keefe, & Yun, 2003). Consequently, Internet and mobile phone use has grown significantly. Table 1 presents data on the size of each country's Internet user population and the level of broadband penetration.

Table 1: Internet user population and broadband penetration in China, Japan and Korea in 2008

	Total number of Internet users (millions)	Internet users (per 100 inhabitants)	Broadband subscribers (per 100 inhabitants)
China	298.0	22.28	6.23
Japan	88.1	69.16	23.65
Korea	37.5	77.83	32.14

Source: International Telecommunication Union [ITU] (2009)

Although China's Internet adoption and broadband subscription rate is the lowest amongst the three countries, it has the largest number of Internet users given the sheer size of its population. Internet use and broadband penetration are discernibly higher in urban areas, with rural areas being served by Internet cafes (Qiu & Zhou, 2005). In Korea, the government has actively encouraged Internet adoption by subsidizing broadband connections and running Internet academies where people can take free lessons on how to use the Internet (John & Gorma, 2002). As for Japan, while it has a lower broadband penetration rate than Korea, this is likely to be due to the fact that it has the highest mobile Internet penetration rate worldwide and there is growing consumer preference for mobile over fixed Internet access (Ishii, 2004).

Internet use amongst youths is on the rise in all three countries. In 2008, 35% of Chinese aged 10-19 were Internet users, while this percentage was below 30 in the previous year (Li & Wang, 2009). In Japan, 68.7% of those aged 6-12 and 94.7% of those aged 13-19 were Internet users in 2007. Compared to the previous year, there was an increase of about 1% in each age group (Ministry of Internal Affairs and Communications, 2008). Notably, amongst young Japanese, the preferred mode of accessing the Internet is via Internet-enabled mobile phones (Boase & Kobayashi, 2008). In Korea, 82.2% of those aged 3-9 and 99.9% of those aged 10-19 were Internet users, up from 79.5% and 99.8% in 2007 respectively (National Internet Development Agency of Korea, 2008). Internet usage is significant even amongst Korean preschoolers, with 58.7% of those aged 3-5 estimated to be users.

Key Concerns

In light of the growing penetration of the Internet amongst youngsters, China, Japan and Korea have introduced various measures to enhance children's online opportunities while minimizing the risks. Our interview findings show that these efforts centre around three main foci: (i) protecting minors from inappropriate and objectionable content; (ii) promoting a more positive, affirming and pro-social online environment through public education and community efforts; (iii) introducing ameliorative measures to stem excessive Internet use and addiction. Each of these topics will be elaborated on in the following sections.

Inappropriate and Objectionable Content

In all three countries studied in this paper, objectionable content is principally dealt with through content filtering although there are differences in how they are operationalised and what they are intended to achieve. China's notorious *Great Firewall* blocks access to online content deemed harmful by the state e.g. pornography and violence, through the use of sophisticated Intranet applications and filtering software (Qiu, 2003). The filtering system blocks content by preventing IP addresses from being routed through and comprises standard firewall and proxy servers at the Internet gateways. Content filtering is a controversial issue for China as its filtering policies have been widely criticized for being politically-motivated and anti-democratic (Lacharite, 2002).[3] Political motivations aside, academics from the Communication University of China confirmed that there is considerable parental pressure for content filtering for the protection of minors, with online pornography being the main issue (personal communication, October 31, 2007). The public distribution of pornography is illegal in China but with growing Internet use, pornography has become significantly easier to access than ever before, thus raising parental concerns (China shuts down 55 more pornography websites, 2009). Besides state-initiated content filtering, ISPs may actually be engaging in a considerable amount of self-censorship based on industry self-regulation codes such as the "Public Pledge of Self-Regulation & Professional Ethics for China Internet Industry" (Klang, 2005). This "self-discipline pact" launched in March 2002 requires ISPs to proactively review the content of both national and international sites, and to then block access to content which is deemed harmful. Internet service and content providers are prepared to sign this pledge in return for access to China's ever-expanding Internet market. ISPs and websites have thus been more careful about adhering to regulations and are pro-actively taking measures to prevent the state from taking action against them (Endeshaw, 2004).

The approach adopted in Japan centres principally on providing content filters for both fixed and mobile Internet access. Responsibility for content filtering has largely been that of the Internet Association Japan (IAJapan) – a non-profit, industry-based organization consisting of about 277 Internet-related organizational and individual members (IAJapan, 2008). The officials at IAJapan and the Ministry of Internal Affairs and

---

3 The debate surrounding the political motivations of China's Internet content filtering policies is beyond the scope of this paper but is the subject of extensive research, e.g. Endeshaw (2004) and Qiu (2003).

Communications (MIC) concurred that the Japanese stance on Internet filtering software is that content filtering is a matter of personal choice and should not be mandated (personal communication, November 6, 2007). However, the MIC has been funding research on the development on filtering software and has also worked with industry groups to raise public awareness about such services. When the Internet was in its incipient stages in Japan, IAJapan developed its own filtering software based on the PICS standard, and distributed it for free to schools and association members, as well as making it available for free download (personal communication, November 6, 2007). Since Japan's Internet market has matured with the emergence of commercial filtering software, IAJapan no longer develops its own version. In addition to filters, IAJapan has also developed the "SafetyOnline2" standard for rating online content which can also be used by commercial filtering software (IAJapan, 2002). Content is classified according to the extent of nudity, sex, violence, and coarse language and each site is graded between 0 and 4, with 0 for "safe" content and 4 for extreme content which is sexually graphic, violent or which is highly defamatory and anti-social. As they surf the Internet, Internet users are encouraged to use these ratings to assess the websites which they encounter, and parents and teachers can also use these ratings to customize the level of filtering which they desire on their own computers. Despite these efforts, a representative of the Institute for HyperNetwork Society observed that the Japanese public did not welcome filtered Internet access (personal communication, November 5, 2007). As of 2008, only 12.9% of households with children aged 18 and below used filtering software on their PCs (Ministry of Internal Affairs and Communications, 2008). In contrast, the same survey found that a significantly higher 21.6% of households used filtering software on the mobile phones. This disparity is not surprising because Japanese youths are avid mobile phone users and parents are increasingly worried that they cannot monitor mobile Internet content which their children access on their personal mobile phones (Ito, 2005). Hence, in 2000, the Japanese government provided free filtering software for mobile phones to address parental concerns. However adoption of this state-provided software gradually declined and commercial companies have since stepped in to offer such services. Indeed, the representative of the Institute for HyperNetwork Society observed that mobile service providers are well aware of parental concerns and have been putting considerable effort into research and development of filters and other methods to ensure a safe mobile Internet environment for minors (personal communication, November 5, 2007). He cited the example of the leading mobile phone operator NTT DoCoMo establishing a "Mobile Society Research Institute" to study the societal impact of mobile phone and mobile Internet use.

In Korea, the protection of minors is a key task of the Korean Internet Safety Commission (KISCOM), a statutory body under the Ministry of Information and Communication. KISCOM promotes self-regulation of Internet and mobile networks by individuals and companies in the interest of fostering a safe online environment. It constantly monitors the circulation of "illegal and harmful content on the Internet", has the authority to define what is deemed "harmful" content and recommends which websites should be blocked countrywide (KISCOM, 2007). It also introduced SafeNet, an Internet content rating standard where content is classified according to the extent of nudity, sex, violence, coarse language and the promotion of the use of drugs, weapons and alcohol, as well as smoking and gambling. Sites are rated between 0 and 4, with 0 for appropriate content and 4 for content which is highly unsuitable for minors (Korea Communications Standards Commission, 2009). KISCOM encourages Internet content creators and providers to self-evaluate and label their content with PICS-standard metatags, and provides parents and schools with filtering software which is compatible with the SafeNet rating system (personal communication, November 2, 2007). In addition to self-rated content, SafeNet develops its own database of "harmful" materials by employing Internet bots which sieve through websites to flag those with inappropriate material. Flagged sites are then manually verified by KISCOM staff, added to the database and rated. The database is updated on a daily basis and KISCOM officials shared that as a testimony to the reliability of the database, at least 60 companies use the Safenet database to create commercial filtering software (personal communication, November 2, 2007).

Promoting a Pro-social Online Environment

Efforts have also been made in each of the three countries to promote pro-social online behaviour so as to foster a more positive Internet environment. This approach is in line with communitarian values of putting the community before self and also the parenting approach of fostering self-control and inculcating positive behaviour in children. In China, the ISC Internet News Service Work Commission set up an online ethics propaganda and education program to promote "civilized online behaviour" (ISC, 2004-2007b). In the case of Japan, IAjapan published an "acceptable code of conduct" for online behaviour which includes such guidelines as:

> When you ask for help, show your appreciation to those who help you. Try to help other people who need help. An online network is like general society, and the same rules apply, including: …
>
> Showing your appreciation to another member who helped you…

Helping out another member who is in trouble

Interactively helping one other (*sic*) [IAJapan, 1996, ¶4-1(2).

The code is published on its website and has been publicized through textbooks issued in children and adult versions. IAjapan also conducts Online Etiquette Accreditation tests for both adults and children to certify their knowledge of the right online etiquette. Certification is issued to those who pass the tests and can be included as a certified skill in employment resumes. Korea's KISCOM runs Cyber Safety Schools at all academic levels with special programs to teach young Koreans how to use the Internet wisely (KISCOM, 2007). It also publishes a quarterly list of websites which contain information that is beneficial for minors.

Apart from imbuing in their citizens the ideal of practicing pro-social online behaviour, the three countries have also sought to make fostering a healthy online environment a community effort. In all three countries, hotlines had been set up for members of the public to report objectionable content.

In China, the public hotline is run by the Internet Society of China (ISC). Formed in 2001 by the Ministry of Information Industry, the ISC is a non-governmental organization which acts as a bridge between government departments and business enterprises which are directly involved in the development of the Internet in China. The ISC's role is foster the development of the Internet in China through academic research and recommendations on industry self-regulation and government policies on the Internet (ISC, 2004-2007a). In June 2004, the ISC founded the China Internet Illegal Information Reporting Centre (CIIRC) which runs a hotline for the public to report objectionable online content. Amongst its missions, it seeks to "Protect the public, minors in particular, from victimization of (*sic*) illegal and harmful information" (CIIRC, no date [a]). While this may include pornography, gambling, violence, extremism or libellous and insulting content (CIIRC, no date [b]), the ISC officials explained that the establishment and maintenance of this hotline was motivated by growing parental concerns about the prevalence of online pornography. The CIIRC acts on public feedback by verifying that the content is indeed objectionable based on China's laws and regulations. It then instructs the Internet service provider (ISP) of the offender to liaise with the website owner to remove the content. For secondary content posted on a website by an external party, the website owner will not be held liable if s/he agrees to remove the offending content. Website owners which refuse to comply are then referred to the Ministry of Public Security for further investigation. At the time of the interview, the CIIRC was handling an average of 20,000 reports a month, most of which related to pornography.

A similar hotline has been in existence in Japan since 2006. It is funded by the National Police Agency and managed by IAJapan. The hotline was established in light of the fact that the "circulation of illegal information on the Internet such as child pornography and advertisements for illegal drugs, and information that is dangerous to public safety and order (harmful information), including information on how to make explosives, contract killings and other illegal activities, has become a problem of our society" (IAJapan, 2006, ¶1). The hotline acts on public feedback about objectionable online material originating from Japan. According to IAJapan officials, the hotline deals with 5,000 cases per month with pornography constituting 44%, child pornography about 24%, substances abuse 15% and fraud and identity theft 14%. Illegal content is reported to the police and in the case of content which is not illegal but considered offensive, IAjapan approaches the content publisher to remove the content, with about 80% of content publishers complying.

In Korea, KISCOM runs the Illegal and Harmful Information Report Centre (SINGO) for Internet users to report any online activity or content which is deemed harmful to minors. The public is thus able to actively participate in the regulation of harmful content on the Internet through reports made via the hotline's website or telephone. KISCOM acts on legitimate reports by asking publishers to remove the offensive content. In addition to the online form, KISCOM also created the "Internet Bluebird" program which Internet users can download from the SINGO website into their own computers (KISCOM, personal communication, November 2, 2007). The Internet Bluebird program expedites public reporting of harmful objectionable content by automating the reporting process. Upon installation, users have to submit their personal particulars. Thereafter, the programme remains dormant on the user's computer but once a user spots any objectionable material, s/he can click on the Internet Bluebird icon to launch a form through which s/he can submit a report to SINGO. KISCOM believes that by automating and simplifying the reporting process, more users will be encouraged to participate in this public reporting exercise. In addition to public complaints (personal communication, November 2, 2007), SINGO also conducts its own monitoring of major portals and user-created content websites to detect objectionable content.

Excessive Usage and Addiction

Excessive Internet use including extremes such as Internet addiction is perceived as another key issue relating to children's Internet safety. Internet addiction is considered to be a growing problem in China and Korea but less so in Japan. Although recent studies have found evidence of Internet and mobile phone addiction amongst some Japanese youths (e.g. Igarashi, Motoyoshi, Takai, & Yoshida, 2008; Sato, 2006) the officials at GLOCOM and IAJapan did not feel that Internet addiction is a particularly serious problem in Japan and confirmed that it is not a focus of government efforts.

In contrast, Internet addiction is an issue of concern in Korea and has been the subject of intense research amongst the country's academic community (see for example Kim *et al.*, 2006, Lee, Oh, Cho, Hong, & Moon, 2001; Lee, Lee, Pail, Kim, & Shin, 2001) and is high on the government's priority list. In 2002, the Korean government set up the Internet Addiction Prevention & Counselling Centre (IAPC) to monitor and tackle the perceived problem of Internet addiction, especially amongst youths (KADO, personal communication, November 1, 2007). The IAPC was established under the auspices of the Korea Agency for Digital Opportunity & Promotion (KADO) and has 45 branch counselling centres located around Korea. As of 2006, they had held 51,777 counselling sessions and had actively promoted Internet addiction prevention and awareness programs. The IAPC also created the K Scale to measure an individual's proneness to Internet addiction (KADO, personal communication, November 1, 2007). This scale is similar to other Internet addiction scales in that it has diagnostic criteria for both self-diagnosis and observer diagnosis. For self-diagnosis, Internet users state their agreement to a series of statements such as "My health has deteriorated due to on-line activities", "I suffer from headaches caused by excessive Internet usage" or "I have failed to follow schedules because of my Internet activities". Similarly, the observer diagnosis instruments contains such statements as "(Individual engages in) Internet usage without meals, breaks, or going to the bathroom", "(Individual) becomes angry and irritated if Internet activity is interrupted" or "(Individual) yells or curses while using the Internet alone". These statements are intended to measure the extent to which a person's Internet use has adversely impacted his/her personal health and ability to function in regular activities. In a study conducted in 2006 using the K Scale, 15.8% of all 16-19-year-olds tested were classified at being at high- or potential-risk for Internet addiction (KADO, personal communication, November 1, 2007). Besides Internet addiction in general, the study found that Korean youths were also at risk of developing mobile phone addiction, gaming addiction, chatting addiction and pornography addiction. The IAPC conducts counselling for people diagnosed with Internet addiction. Besides individual counselling, families of addicts are also counselled because it is believed that the whole family suffers as a result of the addiction. The KADO officials elaborated that some addictions stem from family issues and family counselling thus enables them to initiate appropriate interventions (personal communication, November 1, 2007). For more extreme cases of Internet addiction, the IAPC runs

Internet camps which seek to treat Internet addicts by completely withholding access to computers and making them participate in outdoor activities instead. Consistent with Korean parenting, the camps aim to develop self-control in the participants and to introduce them to healthy alternatives to Internet use. Besides such ameliorative measures, efforts have also been made to introduce preventive measures in the form of public education and campaigns to promote a healthy online gaming culture. This is especially important given the large gaming industry in Korea. In July 2007, several agencies including KADO, the Korea Game Development and Promotion Institute (KOGIA), the Korea Youth Consulting Institute (KYCI) and the Korea Education and Research Information Service (KERIS), signed a memorandum of understanding to work together to create a healthy online culture and to resolve the countries' Internet addiction problem. KOGIA for example conducts research into gaming effects and has also embarked the on the development of educational games to take into account parental concerns.

As for China, the state's response has not been as systematic as that of its Korean counterpart, although there is some evidence that Internet addiction may be a growing mental health problem amongst Chinese adolescents (e.g. Cao & Su, 2006). While Chinese conceptions of Internet addiction are informed by studies of Internet addiction in other countries (e.g. Young, 1998), Chinese studies have developed their own definitions of what constitutes Internet addiction. For example, the academics we interviewed explained that the Chinese define an Internet addict as someone who uses the Internet excessively for a single purpose such as gaming or pornography, while a non-addict might also be a heavy Internet user, but would use it for a variety of purposes (Communication University of China, personal communication, October 31, 2010). These academics also felt strongly that China needs to step up public education efforts to teach people to manage their use of the Internet so that they are not fixated on any single aspect of the Internet such that they neglect their responsibilities. As with Korea, China has also experimented with Internet boot camps to treat addiction. Over 200 organisations, some of which are government-funded, are offering treatment for Internet addiction (Graham-Harrison, 2009). Many of them adopt a military camp approach where addicts are made to give up the Internet "cold turkey", forced to reside in a military camp and engage in physical activities to cure themselves of the addiction (Ransom, 2007).[4] Counselling for the addicts and their families may also be conducted.

4 In a recent incident in August 2009, a Chinese teen died hours after checking into an Internet bootcamp, apparently from abuse by the staff. His death has been widely publicized in the media and has prompted a government investigation. See Graham-Harrison (2009).

The academics we spoke to cautioned that while such boot camps have proliferated nation-wide, the efficacy of their treatment methods have not been systematically examined and warranted closer scrutiny and regulation (Communication University of China, personal communication, October 31, 2010).

Apart from these state-initiated efforts at stemming Internet addiction, media companies in China and Korea are also responding to these concerns about excessive Internet use with service innovations. This is understandable as the sizable online youth populations in these countries give local media companies the impetus to create web products and services which can appeal to young children and their parents. For example, Korea's top online portal and search engine, Naver, set up the Jr Naver search engine to serve content tailored for the young. It has a child-friendly interface and age-appropriate content such as nursery rhyme flash videos. The representative of NHN Corporation, Jr Naver's parent company, explained that it even has a "homework helper" service to cater to Korean society's emphasis on academic pursuits (personal communication, November 1, 2007). Another interesting example can be found in China's Mole Manor, the leading virtual world and game website for children aged 7-12. The website reminds players to take a break every 45 minutes and is shut down from 12 pm-6 am daily in recognition of children's poor ability to regulate their online activity (Wen, 2009).

Conclusion

This study's findings show that while China, Japan and Korea seek to boost Internet adoption, the social impact of the Internet, especially on minors, is an issue of societal concern and the target of policy interventions, non-governmental initiatives and industry responses. These measures to promote Internet safety for children are reflective of and shaped by the three countries' cultural values. The concerted efforts which have been made in developing their own standards for rating online content (SafetyOnline2 in Japan and SafeNet in Korea) and domestic innovations in developing content filtering software for both fixed and mobile devices reflect societal interest in these countries to shield children from adverse influences in their external environment. This approach resonates with the protective parenting style that is salient in the three countries (Gao *et al.*, 1996; White, 1993; Wu, 1996a; 1996b), and where childhood is considered a critical period for nurturing the child (Chao & Tseng, 2002). The level of family and societal investment in each child's development, especially their educational performance, also demands that children are prevented from being corrupted by unsavoury online influences that can be detrimental to their academic performance. Hence, the filtering of objectionable content has been a clear focus of regulatory initiatives aimed at improving children's online experiences. The concern about excessive Internet use and addiction can also be traced to these societies' strong emphasis on children's academic achievements and discipline. Industry innovations such as Naver's homework helper service in Korea and the service shutdown for China's Mole Manor are clear responses to the preoccupation with children's academic pursuits and their possible displacement by online activities. Children are deemed to have to be disciplined and to develop their self-control, with light corporal punishment and the modelling of good behaviour by parents used to instil this quality in children (Hendry, 1986; Kim *et al.*, 1997; Machida, 1996). This concern with self-control can also be discerned in the extreme boot camp approach taken to "treat" cases of Internet addiction. Kwee, Komuro-Venovic and Kwee (2009) argue that the appeal of the boot camp approach in China and Korea may lie in the fact that it involves discipline and corrective action, thus resonating with these societies' authoritarian leanings and emphasis on discipline. Another approach geared at improving children's online experiences is the push to promote a pro-social online environment. Quite apart from shielding the children from negativity online, the three countries have sought to inculcate a more positive and healthier online environment by promulgating codes for online etiquette. Such codes of proper conduct are very much in line with these societies which have traditionally stressed hierarchy and obedience (Tu, 1996), although individual conformance to such codes is definitely not guaranteed even in such circumstances. The active use of hotlines is another reflection of these societies' emphasis on hierarchy and communitarianism, where everyone is encouraged to play a part in developing a healthy online environment by reporting objectionable content, and partaking in this sense of a shared mission. Despite the apparent congruence of these regulatory initiatives with these countries' cultural values and societal priorities, it remains to be seen whether these initiatives to improve children's online experiences are perceived by parents, children and educators as effective.

It behoves me therefore to outline several limitations to this study which future research by this author will seek to address. First, although a range of governmental, non-governmental and industry organisations were interviewed, more emphasis was placed on governmental organisations. Consequently, alternative voices on the issue of children's Internet safety, such as those of parent, teacher and youth organisations have not been reflected. Second, given the range of topics which the interviews had to cover within the constraints of time, we were unable to probe further about the efficacy and public perceptions of these efforts at promoting children's Internet safety. Third, there were several issues relating to children's Internet safety which were unique to each country that were not covered during the interviews. These issues warrant academic attention and urgent policy responses, and they include the growing use of online dating websites for teen prostitution and sexual victimisation in Japan and Korea (Kioka, 2003; McCoy, 2003), the rising trend of Internet suicide pacts in Japan (Naito, 2007) and the lack of supervision of children using Internet cafes in China (Qiu & Zhou, 2005).

References

Anagnost, A. (1997). Children and National Transcendence in China. In K.G. Lieberthal, S.F. Lim, & E.P. Young (eds.), *Constructing China: The Interaction of Culture and Economics* (pp. 195-222). Ann Arbour, MI: Centre for Chinese Studies, University of Michigan.

Bleha, T. (2005). Down to the Wire. *Foreign Affairs.* May/June 2005. Retrieved Augustus 4, 2009, from http://fc.xubin.googlepages.com/DowntotheWire.pdf.

Boase, J. & Kobayashi, T. (2008). Kei-Tying Teens: Using Mobile Phone E-mail to Bond, Bridge, and Break with Social Ties – A Study of Japanese Adolescents. *International Journal of Human-Computer Studies*, *66*, 930-943.

Cao, F. & Su, L. (2006). Internet addiction among Chinese adolescents: Prevalence and psychological features. *Child: Care, Health and Development*, *33*, 275-281.

Chang, K.-S. (2007, April). *Politics of the Social Investment Family: Education and State-Family Relations in South Korea*. Conference on the Changing Asian Family as a Site of the State Politics at the Asia Research Institute, Singapore.

Chao, R. & Tseng, V. (2002). Parenting of Asians. In M.H. Bornstein (ed.), *Handbook of Parenting* (2nd ed., Vol. 4, pp. 59-93). Mahwah N.J.: Lawrence Erlbaum Associates.

Chee, B.W.L. (2000). Eating Snacks and Biting Pressure: Only Children in Beijing. In J. Jun (ed.), *Feeding China's Little Emperors: Food, Children and Social Change* (pp. 48-70). Stanford, CA: Stanford University Press.

Chen, C.S. & Stevenson, H.W. (1995). Motivation and Mathematics Achievement: A Comparative Study of Asian-American, Caucasian-American, and East Asian High School Students. *Child Development*, *66*, 1215-1234.

Chen, C.S., Lee, S.Y., & Stevenson, H.W. (1996). Academic Achievement and Motivation of Chinese Students: A Cross-national Perspective. In S. Lau (ed.), *Growing Up the Chinese Way* (pp. 69-92). Hong Kong: The Chinese University Press.

Chen, S.J. (1996). Are Japanese Young Children Among the Gods?, In D.W. Shwalb & B.J. Schwalb (eds.), *Japanese Childrearing: Two Generations of Scholarship.* (pp. 31-43). New York: Guilford Press.

China Shuts Down 55 More Pornography Websites. (2009, January 28). *China Daily*. Retrieved July 10, 2009, from http://www.chinadaily.com.cn/china/2009-01/28/content_7430510.htm.

CIIRC. (no date [a]). *Mission.* Retrieved July 27, 2009, from http://ciirc.china.cn/about/txt/2006-06/08/content_124006.htm.

CIIRC. (no date [b]). *China Internet Illegal Information Reporting Centre*. Retrieved July 27, 2009, from http://ciirc.china.cn.

Connor, M.E. (2002). *The Koreas: A Global Studies Handbook.* Santa Barbara, CA: ABC-CLIO Inc.

Cotton, J. & van Leest, K.H. (1996). The New Rich and the New Middle Class in South Korea: The Rise and Fall of the "Golf Republic". In R. Robison & D.S.G. Goodman (eds.), *The New Rich in Asia: Mobile Phones, McDonalds and Middle-Class Revolution* (pp. 185-203). London: Routledge.

Cui, J. (2009, July 22). Green Dam Filtering into Computers. *China Daily*. Retrieved July 10, 2009, from http://www.chinadaily.com.cn/bizchina/2009-07/22/content_8457090.htm.

Endeshaw, A. (2004). Internet Regulation in China: The Never-ending Cat and Mouse Game, *Information & Communications Technology Law, 13* (1), 41-57.

Gao, G., Ting-Toomey, S., & Gudykunst, W.B. (1996). Chinese Communication Processes. In M.H. Bond (ed.), *The Handbook of Chinese Psychology* (pp. 280-293). Hong Kong: Oxford University Press.

Graham-Harrison, E. (2009, Aug 6). China Probes Internet Bootcamp after Teen Dies. *Reuters*. Retrieved August 7, 2009, from http://www.reuters.com/article/technologyNews/idUSTRE5750K520090806.

Gu, J. (1990). Household Structure in Contemporary China. In Y. Zeng, C. Zhang, & S. Peng (eds.), *Changing Family Structure and Population Aging in China* (pp. 92-106). Beijing: Peking University Press.

Hau, K.T. & Salili, F. (1996). Achievement Goals and Causal Attributions of Chinese Students. In S. Lau (ed.), *Growing Up the Chinese Way* (pp. 121-146). Hong Kong: The Chinese University Press.

Hendry, J. (1986). *Becoming Japanese: The World of the Pre-school Child.* Manchester: Manchester University Press.

Hendry, J. (2003). *Understanding Japanese Society* (3rd ed.). London: Routledge.

Ho, D.Y.F. (1986). Chinese Patterns of Socialisation: A Critical Review. In M.H. Bond (ed.), *The Psychology of the Chinese People* (pp. 1-37). Hong Kong: Oxford University Press.

Holloway, S.D., Suzuki, S., Yamamoto, Y., & Mindnich, J.D. (2006). Relation of Maternal Role Concepts to Parenting, Employment Choices, and Life Satisfaction Among Japanese Women. *Sex Roles*, *54* (3/4), 235-249.

IAJapan. (1996). *Recommended Etiquette for Online Service Users*. Retrieved July 27, 2009, from http://www.iajapan.org/en/rule/etiquette.html.

IAJapan. (2002). *Internet Association, Japan Releases New Rating Standards for Rating Harmful Content on the Internet and Offers Upgraded Filtering System Applying the New Standards.* Retrieved July 27, 2009, from http://www.iajapan.org/rating/press/sfs301-press-en.html.

IAJapan. (2006). *Operational Launch of the Internet Hotline Centre – An Integrated Point for Reporting Illegal or Harmful Information to Enable Safe Use of the Internet.* Retrieved July 27, 2009, from http://www.Internethotline.jp/index_e.html.

IAJapan. (2008). *Internet Association Japan.* Retrieved July 27, 2009, from http://www.iajapan.org/index-en.html.

Igarashi, T., Motoyoshi, T., Takai, J., & Yoshida, T. (2008). No Mobile, No Life: Self-perception and Text-message Dependency Among Japanese High School Students. *Computers in Human Behaviour, 24*, 2311-2324.

ISC. (2004-2007a). *Internet Society of China.* Retrieved July 27, 2009, from http://www.isc.org.cn/isc_eIntroduction/index.htm.

ISC. (2004-2007b). *Memorabilia.* Retrieved July 27, 2009, from http://www.isc.org.cn/isc_eProvinceSociety/index.htm.

Ishii, K. (2004). Internet Use via Mobile Phone in Japan. *Telecommunications Policy, 28*, 43-58.

Ito, M. (2005). Mobile Phones, Japanese Youth, and the Re-Placement of Social Contact. In R. Ling & P.E. Pedersen (eds.), *Mobile Communications: Re-negotiation of the Social Sphere* (pp. 131-148). London: Springer.

ITU. (2009) *Internet Indicators: Subscribers, Users and Broadband Subscribers.* Retrieved August 5, 2009, from http://www.itu.int/ITU-D/icteye/Reporting/ShowReportFrame.aspx?ReportName=/WTI/Information TechnologyPublic&RP_intYear=2008&RP_intLanguageID=1.

John, Y.J. & Gorma, G.E. (2002). Internet Use in South Korea. *Online Information Review, 26* (5), 335-344.

Kim, K., Ryu, E., Chon, M.-Y., Yeun, E.J., Choi, S.-Y., Seo, J.-S. *et al.*. (2006). Internet Addiction in Korean Adolescents and Its Relation to Depression and Suicidal Ideation: A Questionnaire Survey. *International Journal of Nursing Studies, 43*, 185-192.

Kim, M.K. (1994). Achievement Motivation and Individualism in Korean Adolescents. In G. Yoon, & S.-C. Choi (eds.), *Psychology of the Korean People: Collectivism and Individualism* (pp. 265-267). Seoul: Dong-A Publishing and Printing Co., Ltd.

Kim, U., Park, Y.-S., Kwon, Y.-E., & Koo, J. (2005). Values of Children, Parent-Child Relationship, and Social Change in Korea: Indigenous, Cultural and Psychological Analysis. *Applied Psychology: An International Review, 54*(3), 338-354.

Kim, W.J., Kim, L.I., & Rue, D.S. (1997). Korean American Children. In G. Johnson-Powell & J. Yamamoto (eds.), *Transcultural Child Development* (pp. 183-207). New York: Wiley.

Kioka, Y. (2003). *Dating Sites and the Japanese Experience*. Proceedings of the Experts Meeting of Children, Mobile Phones and the Internet, Tokyo, Japan. Retrieved July 20, 2009, from http://www.iajapan.org/hotline/mobilepdf/4_KIOKA.pdf.

KISCOM. (2007). *Making a Safer and Sounder Cyberspace with KISCOM*. Seoul: Kiscom.

Korea Communications Standards Commission. (2009). *Rating Standards*. Retrieved July 27, 2009, from http://www.safenet.ne.kr/english/introduction/rating_standards.html.

Klang, M. (2005). *Controlling Online Information: Censorship & Cultural Protection*. A Presentation Given at "WSIS, Internet Governance and Human Rights" Uppsala, October 3, 2005.

Kwee, A.W., Komuro-Venovic, E., & Kwee, J.L. (2009). Treatment Implications from Etiological and Diagnostic Considerations of Internet Addiction: Cautions with the Boot Camp Approach. *Proceedings of the e-CASE 2009 International Conference* (pp. 1459-1476), Singapore, January 8-10, 2009.

Lacharite, J. (2002). Electronic Decentralisation in China: A Critical Analysis of Internet Filtering Policies in the People's Republic of China. *Australian Journal of Political Science, 37* (2), 333-346.

Lee, H., O'Keefe, R.M., & Yun, K. (2003). The Growth of Broadband and Electronic Commerce in South Korea: Contributing Factors. *The Information Society, 19*, 81-93.

Lee, M.S., Oh, E.Y., Cho, S.M., Hong, M.J., & Moon, J.S. (2001). An Assessment of Adolescent Internet Addiction Problems Related to Depression, Social Anxiety and Peer Relationship. *Journal of the Korean Neuropsychiatry Association, 40* (4), 616-626.

Lee, S.B., Lee, K.K., Pail, K.C., Kim, H.W., Shin, S.K. (2001). Relationship Between Internet Addiction and Anxiety, Depression, and Self Efficacy in Middle and High School Students. *Journal of the Korean Neuropsychiatry Association, 40* (6), 1174-1184

Lee, Y.G. & Huh, E. (2004). Consumption and Saving Behaviour of Older and Younger Baby Boomers in Korea. *Journal of Family and Economic Issues, 25*, 507-26.

Li, W. & Wang, X. (2009, Jan 13). CNNIC: China had 298 million Netizens by Dec 2008. *China Daily*. Retrieved July 10, 2009, from http://www.chinadaily.com.cn/bizchina/2009-01/13/content_7392547.htm.

Lim, S.S. (2008). Technology Domestication in the Asian Homestead: Comparing the Experiences of Middle Class Families in China and South Korea. *East Asian Science, Technology and Society, 2*(2), 189-209.

MacDonald, D.S. (1996). *The Koreans: Contemporary Politics and Society* (3rd ed.). Boulder, CO: Westview Press.

Machida, S. (1996). Maternal and Cultural Socialization for Learning: Lessons Learned and Prospects Ahead. In D.W. Shwalb & B.J. Schwalb (eds.), *Japa-*

nese Childrearing: Two Generations of Scholarship (pp. 241-259). New York: Guilford Press.

McCoy, A. (2003). Blaming Children for Their Own Exploitation: The Situation in East Asia. *ECPAT 7th Report on the Implementation of the Agenda for Action Against the Commercial Sexual Exploitation of Children*. Retrieved July 15, 2009, from http://www.make-it-safe.net/esp/pdf/Blaming_Children.pdf.

Ministry of Internal Affairs and Communications. (2008). *Communications Usage Trend Survey in 2007 Compiled*. Retrieved July 10, 2009, from http://www.johotsusintokei.soumu.go.jp/tsusin_riyou/data/eng_tsusin_riyou02_2007.pdf.

Naito, A. (2007). Internet Suicide in Japan: Implications for Child and Adolescent Mental Health. *Clinical Child Psychology and Psychiatry*, *12* (4), 583-597.

National Internet Development Agency of Korea. (2008). *Survey on the Internet Usage*. Retrieved July 20, 2009, from http://isis.nida.or.kr.

Qiu, J. & Zhou, L. (2005). Through the Prism of the Internet Café: Managing Access in an Ecology of Games. *China Information*, *19*, 261-296.

Qiu, J.L. (2003). *The Internet in China: Data and Issues*. Working Paper Prepared for Annenberg Research Seminar on International Communication. 1 October 2003.

Ransom, I. (2007, March 12). Chinese Boot Camps Tackle Internet Addiction. *New York Times*. Retrieved August 5, 2009, from http://www.nytimes.com/2007/03/12/technology/12iht-addicts.4880894.html?_r=1.

Sato, T. (2006). Internet Addiction Among Students: Prevalence and Psychological Problems in Japan. *Japan Medical Association Journal*, *49*(7/8), 279-283.

Stevenson, H.W. & Lee, S.-Y. (1996). The Academic Achievement of Chinese Students. In M.H. Bond (ed.), *The Handbook of Chinese Psychology* (pp. 124-142). Hong Kong: Oxford University Press.

Tu, W.-M. (ed.) (1996). *Confucian Traditions in East Asian Modernity: Moral Education and Economic Culture in Japan and the Four Mini-Dragons*. Cambridge, MA: Harvard University Press.

Vaughn, C.A. (1996). Socialisation and School Adaptation: On the Lifework of George De Vos. In D.W. Shwalb & B.J. Schwalb (eds.), *Japanese Childrearing: Two Generations of Scholarship* (pp. 85-96). New York: Guilford Press.

Wang, Y. & Cui, N. (2004, March 26). Boom of After-school Education in China. *China Daily*. Retrieved July 20, 2009, from http://www.chinadaily.com.cn/english/doc/2004-03/26/content_318433.htm.

Wen, Y. (2009, May 5). Totally Wrong to Bundle Online Ggame. *The Global Times*. Retrieved July 20, 2009, from http://china.globaltimes.cn/China News/2009-05/429221.html.

White, M. (1993). *The Material Child: Coming of Age in Japan and America*. New York: Free Press.

Wu, D.Y.H. (1996a). Chinese Childhood Socialisation. In M.H. Bond (ed.) *The Handbook of Chinese Psychology* (pp. 143-154). Hong Kong: Oxford University Press.

Wu, D.Y.H. (1996b). Parental Control: Psychocultural Interpretations of Chinese Patterns of Socialization. In S. Lau (ed.), *Growing Up the Chinese Way* (pp. 1-28). Hong Kong: The Chinese University Press.

Young, K.S. (1998). Internet Addiction: The Emergence of a New Clinical Disorder. *CyberPsychology and Behavior, 1* (3), 237-244.

Appendix – List of Organisations Interviewed

China

China Internet Network Information Centre

Chinese Academy of Social Sciences

Communication University of China

EU-China Information Society Project

Internet Society of China

State Council Information Office

Tsinghua University

Japan

Centre for Global Communications (GLOCOM)

Institute for HyperNetwork Society

Internet Association of Japan (IAjapan)

Keio University

Ministry of Economy, Trade and Industry

Ministry of Internal Affairs and Communications

Nikkei Media Lab

Korea

Centre for Internet Addiction, Korea Agency for Digital Opportunity & Promotion (KADO)

Chung-Ang University

Korea Broadcasting Commission (KBC)

Korea Game Industry Agency (KOGIA)

Korea Internet Safety Commission (KISCOM)

NHN Corporation

OhmyNews International

Contributors

Núria Araüna is a doctoral candidate and fellow researcher at the Communication Studies Department at Rovira i Virgili University. Her research interests are gender representations in audiovisual formats such as music videos, TV series and social networking sites. She is currently involved in a state-funded research project looking at the relationship between media constructions and violence towards women in Spain and in a collaborative video project with women.

Joke Bauwens, PhD, is professor of Media Sociology and Media and Culture at the Department of Media and Communication Studies, Free University Brussels (Vrije Universiteit Brussel). Her research activities, situated in the research group IBBT-SMIT, involve social and cultural practices in digital and virtual media-contexts. She publishes on the following topics: young people and Internet, morality in social virtual worlds, digital television, media literacy and digital arts. She is Belgian team member of EU Kids Online.

Périne Brotcorne is researcher at the FTU Fondation Travail-Université (Brussels/Namur) since 2007. She has a master degree in history and in sociology (ULB). She worked at the Research Group of Information and Humanities at the ULB where she conducted research on the use of Internet by students at the University. She also collaborated in several projects at the Work & Technology Research Centre of FTU (Namur) on the topic of digital divide and Internet education.

Cédric Courtois works as a junior researcher at the research group for Media and ICT (MICT) at Ghent University. His current research interests fall under the broad umbrella of media consumption in a converging media landscape, youth and new media and online prosuming.

Lucrezia Crescenzi, PhD, is a postdoctoral researcher at the Interactive Media Lab of the University of Barcelona. She graduated in Developmental Psychology at the La Sapienza University in 2006 and she holds a PhD in Communication, Art and Education from University of Barcelona (2010). Lucrezia Crescenzi is a specialist in content analysis methodology and has been involved in the development of projects about screens and childhood, media literacy, diversity and social inclusion in media contexts, and mobile-learning.

Rozane De Cock (PhD Social Sciences 2007) is professor at the Catholic University of Leuven (Belgium), a member of the Centre for Media Culture and Communication Technology (KULeuven) and the

Brussels Centre of Journalism Studies (HUBrussel). Her research focuses on children and their news consumption and the interdisciplinary field of journalism, justice and law (crime reporting). She teaches courses such as News Production and News Reception; Qualitative Methods of Communication Research; News and News Effects.

Yvonne A.W. De Kort, PhD, is associate professor of environmental psychology and co-director of the Game Experience Lab at Eindhoven University of Technology. Her main research interests are restorative effects of environments, situated social interaction and play (digital gaming), and locatedness in mediated communication and social interaction.

Nayla Fawzi is a research and teaching assistant at the Institute for Communication Studies and Media Research of Ludwig-Maximilians-University Munich in Germany. She studied communication science, educational and business studies at the University of Mainz and Dijon (France). Her research focuses on political communication (e.g. mediatisation of politics) and online communication in particular cyberbullying.

Bernhard Goodwin is working as researcher, teacher and department manager at the Institute for Communication Science and Media Research of the LMU in Munich in Germany. He studied communication science, law and computer science at the LMU and at the UCT in Cape Town, South Africa. As graduate student he specialised in the field of science communication at the TUM in Munich, with a focus on the role of scientists in public communication of their research. Upon finishing his PhD research project he returned to LMU and does research about lay communicators there.

Wannes Heirman is a research and teaching assistant at the Department of Communication Studies of the University of Antwerp. After obtaining his master's degree in communication sciences (University of Antwerp, 2006), he started doing research work for the research team MIOS. In the period 2008-2009 he worked on a research report concerning cyberbullying. The main focus of his PhD-research is devoted to the question how young children and teenagers deal with their privacy on the Internet.

Wijnand A. IJsselsteijn, PhD, has a background in psychology and artificial intelligence, with an MSc in cognitive neuropsychology from Utrecht University, and a PhD in media psychology/HCI from Eindhoven University of Technology. He currently holds a position as associate professor within the Human-Technology Interaction Group at the Eindhoven University of Technology. His focus is on digital social media, immersive media technology and embodiment and gaming.

Veronika Kalmus, PhD, is Professor of Media Studies at the Institute of Journalism and Communication at the University of Tartu in Estonia. Her research interests are socialisation, inter-generational relationships, and changes in human values. She leads the research project "Generations and inter-generational relationships in the emerging information society", financed by the Estonian Science Foundation, and participates in several other projects, including EU Kids Online.

Ellen M. Kraft, PhD, is an Assistant Professor in the School of Business at Richard Stockton College of New Jersey. She received her B.S.I.E. from Lehigh University, M.S.E. from West Virginia University, and PhD from Auburn University. Her research about adolescent and college student cyberbullying has been published in the Journal of Techno-ethics, Journal of Cyber-criminology, WIT Transactions on Information and Communications Technology, and in Truths and Myths of Cyber-bullying: International Perspectives on Stakeholder Responsibility and Children's' Safety (Peter Lang, 2010).

Eva Lievens, PhD, obtained a law degree at the University of Ghent and a Masters degree in Transnational Communications and Global Media at Goldsmiths College, London. Eva has been a member of the Interdisciplinary Centre for Law & ICT (K.U.Leuven) since 2003 and obtained her PhD in law in June 2009. Since October 2011 she is a postdoctoral Fellow of the Research Fund Flanders. Her research deals with legal challenges posed by new communication phenomena, such as the regulation of audiovisual media services, user-generated content, and social networks.

Sun Sun Lim, PhD, is Associate Professor and Deputy Head of the Department of Communications and New Media, National University of Singapore. She studies technology domestication by youths and families in Asia and has conducted fieldwork in China, Japan, Korea and Singapore. She has lectured and published widely, including in the *Journal of Computer Mediated Communication*, *New Media & Society*, *Communications of the ACM*, and *Telematics & Informatics*. She is a member of Singapore's National Youth Council which advises on the country's youth-centred policies.

Peter Mechant works as a researcher at the research group for Media and ICT (MICT), Ghent University. He focuses his research on topics such as social software, folksonomies, online communities and Web 2.0-websites.

Sara Mels holds a master's degree in Modern History and in International Relations and Conflict Resolution from the K.U.Leuven. She is currently project coordinator at the University Centre Saint Ignatius Antwerp.

Willem Oosting has a background in both technology- and social sciences, with a master's degree in Human Technology Interaction. Throughout his career he was involved in various interdisciplinary research and development projects in the field of "Human Computer Interaction". After developing telemedicine systems as a usability engineer, he studied the fundaments of parental mediation of children's digital gaming as an academic. Currently he works in a business-driven context on automation of knowledge dissipation using CMS. In general terms; his career revolves around research and development of ICT, while maintaining a user-centered perspective.

Steve Paulussen, PhD, is senior researcher at the IBBT research group for Media & ICT (IBBT-MICT) at Ghent University (Belgium), where he is involved in different projects that relate to issues of media production and use in a changing digital environment. Currently, he is involved in an IWT funded strategic research project on "User Empowerment in a Social Media Culture" (EMSOC). He is also part-time lecturer in journalism and new media at the University of Antwerp and the Vrije Universiteit Brussel (VUB).

Triin Roosalu is researcher and lecturer in the Institute for International and Social Studies, Tallinn University, Estonia. With a degree in economics and in European Studies from Tartu University, she is preparing her dissertation in sociology in Tallinn University on balancing caring and careering in post-socialist countries. Over the last five years she has published on issues of parenthood, working time, labour market attachment, and lifelong learning. She has co-edited a book *Learning in Transition: Policies and Practices of Lifelong Learning in Post-Socialist Countries* and is currently preparing an edited volume on women's employment in post-socialist Europe.

Itxaso Tellado is a lecturer in the Department of Pedagogy at the University of Vic. She holds a doctorate in education and a Certificate of Women's Studies by Northern Illinois University. Her research interests are adult education, cultural minorities and gender issues. She is currently involved in two state-funded research projects on gender issues. She is member of CREA, Center of Research in Theories and Practices that Overcome Inequalities. She participated in an applied project on bullying in schools under the Daphne II Programme of the European Commission.

Peter Thijssen, PhD, is Associate Professor in Political Sociology and Methodology in the Department of Political Science at the University of Antwerp (Belgium). He is member of the research group Media, Movements & Politics (M^2P). His current research focuses on political participation and intergenerational solidarity.

Christiane Timmerman holds a PhD in social and cultural anthropology. She is director of the Centre for Migration and Intercultural Studies of the University of Antwerp and academic director of the University Centre Saint-Ignatius Antwerp.

Iolanda Tortajada, PhD, is a senior lecturer at the Communication Studies Department, Rovira i Virgili University. Her research interest lies in the field of inequalities of access and representation to media, specifically regarding ethnic minorities and gender issues. Currently she is coordinating a basic research project sponsored by Spain's Ministry of Science and Innovation named "Media and gender violence: Do they encourage it or do they prevent it?" which deals with reception studies, SNS sites analysis and discourse analysis of TV series and music videos.

Heidi Vandebosch, PhD, is an associate professor at the Department of Communication Studies at the University of Antwerp (Belgium), specialising in media sociology and health communication. She has an MA in Communication Studies (K.U.Leuven) and a PhD in Social Sciences (K.U.Leuven). Her research focuses on media/ICT uses and effects. She has published internationally about: cyberbullying amongst youngsters, TV news consumption of adolescents, TV and ICT use of elderly, TV use and obesity among pre-school children, media consumption patterns among prisoners, TV game show viewers, effects of mass media on name-giving and weather and lunar effects on television viewing.

Chris Vleugels is a sociologist (Katholieke Universiteit Leuven) who is working since 2006 as a researcher at IBBT-SMIT, Vrije Universiteit Brussel on interdisciplinary projects regarding new and emerging technologies and applications within the cultural field. His main research interests include e-culture, e-participation, media literacy, sociology of youth and qualitative research.

Michel Walrave, PhD, is an associate professor at the University of Antwerp, where he is responsible for the research group MIOS and head of the Department of Communication Studies. He holds a master's degree in Communication Studies, a master in Information Science and a PhD in Social Sciences. His research is situated in the area of societal implications of ICT in general, and ICT-use related risks in particular. He has conducted several national and international research projects on adolescents' ICT uses and related risks (online self-disclosure, cyberbullying), e-marketing and e-privacy (online marketing and data mining, social network sites and privacy). He lectures on societal implications of ICT, governmental and social profit communication, e-marketing and privacy.

Jinchang Wang received his PhD in operations research and information technology at Georgia Institute of Technology in 1990. He is currently a Professor in Richard Stockton College of New Jersey, USA. His research papers have been published in the following journals *Mathematical Programming, ORSA Journal on Computing, Annals of Mathematics and Artificial Intelligence, Journals of Automated Reasoning, International Journal of Approximate Reasoning, Mathematical and Computer Modelling, Knowledge-based Systems, Computers and Operations Research, Journal of Combinatorial Optimization, International Journal of Techno-ethics, International Journal of Cyber Criminology* etc.

Cilia Willem, PhD, is a lecturer and researcher at the University of Barcelona, where she has taught digital media for the last seven years at the Audiovisual Communication Studies. Her research interest lies in young adults' identity construction in self-generated media, and the role of digital literacy in young people's lives. She is a European project manager at the Interactive Media Lab (LMI) and is currently involved in a state-funded research project looking at the relationship between media constructions and violence towards women in Spain.